Biology Labs That Work:
The Best of How-To-Do-Its

Edited by Randy Moore

Published by the National Association of Biology Teachers (NABT), 11250 Roger Bacon Drive #19, Reston, Virginia 22090-5202

ISBN 0-941212-18-1

Copyright © 1994 by the National Association of Biology Teachers

All rights reserved. The laboratory exercises contained in this book may be reproduced for classroom use only. This book may not be reproduced in its entirety by any mechanical, photographic or electronic process or in the form of a photographic recording, nor may it be stored in a retrieval system, transmitted or otherwise copied for any other use without written permission of the publisher.

Cover photo of a Randolph-Macon College student doing lab work was provide courtesy of photographer Bill Denison.

Printed in the United States of America by Lancaster Press, Inc., Lancaster, Pennsylvania.

While every effort was made to anticipate questions and situations that could arise, the safe implementation of these activities must depend on the good judgment of teachers and is the responsibility of the local school district/institution. NABT recognizes the pervasive social phenomenon of litigation with respect to even the most unfounded claims. For this reason, NABT disclaims any legal liability for claims arising from use of these experiments. This information has been provided to teachers and to schools as a service to the profession; and NABT provides this material only on the basis that NABT has no liability with respect to its use.

NABT believes that under the guidance of a properly trained and responsible teacher, these experiments can be safely conducted in the classroom or laboratory. Before conducting these activities, NABT recommends that teachers check with their science supervisors, district or state education offices, and/or their local authorities for general lab safety and disposal guidelines. For further information on working safely with microorganisms, refer to NABT's publication, *Working with DNA & Bacteria in Precollege Science Classrooms*. Also, for more information on the safe handling and disposal of microorganisms, consult the current *Flinn Chemical Catalog/Reference Manual*, Flinn Scientific, Inc., Batavia, Illinois.

Preface

Since its inception, the *How-To-Do-It* section of *The American Biology Teacher* has been immensely popular. Requests for copies of articles come in from all over the world, and teachers regularly tell me that they use *How-To-Do-It* articles in their classes. The success of these articles is due to their quality and the fact that the best teachers know that biology, like any science, is a process—that is, it is something that we *do*. In this sense, *Biology Labs That Work: The Best of How-To-Do-Its* is a guide for how to *do* biology.

The articles in this book were chosen for their safety, simplicity, dependability, economy and diversity. Each activity can be used alone or as a starting point for helping students design follow-up experiments for an in-depth study of a particular topic. Regardless of their use, you can depend on these experiments to help you teach biology as biology is done—that is by training students to make keen observations, form hypotheses, design experiments, interpret data, and communicate their results and conclusions.

I thank the authors of these articles for letting me include updated versions of their work in this book. I also thank Michele Bedsaul, Chris Chantry and Katherine Munson for their help with this book, the members of NABT's Publications Committee for helping me select these articles, and NABT for giving me the opportunity to produce this book. I hope you like it.

Randy Moore

Contents

Cell and Molecular Biology

Genetic Transformation of Bacteria	8
A Simple Demonstration of Fermentation	11
Demonstrating the Effects of Stress on Cellular Membranes	14
Demonstrating Osmosis and Anthocyanins Using Purple Onion	19
pH and Rate of Enzymatic Reactions	22
A Simple DNA Isolation Technique Using Halophilic Bacteria	25
A Paper Model of DNA Structure and Replication	28

Microbes and Fungi

Pasteurized Milk as an Ecological System for Bacteria	32
Slime Molds in the Laboratory: Moist Chamber Cultures	41
The Almost Ideal Lab--Mutualistic Nitrogen Fixation	47
Quantifying Intracellular Water Regulation in a Single-Celled Organism	58
Using Yeast and Ultraviolet Radiation To Introduce the Scientific Method	65
pH and Microbial Growth	69

Plants

Accurately Measuring Transpiration	76
The Influence of pH on the Color of Anthocyanins and Betalains	77
Plant Eco-Physiology: Experiments on CAM Using Minimal Equipment	81
Using Dandelion Flower Stalks for Gravitropic Studies	90
Thin Layer Chromatography (TLC) of Chlorophyll Pigments	98
Rapid Germination of Pollen *In Vitro*	104
Some Plant Hormone Investigations That Work	107

Animals

Artificial Urine for Laboratory Testing	114
The Sea Urchin Embryo: A Remarkable Classroom Tool	118
Artificial Urine Test To Simulate the Test for Pregnancy	125
Disease Detective: A Game Simulation of a Food Poisoning Investigation	128

Evolution and Ecology

Economics and Biology: An Analogy for the Presentation of the Niche Concept	136
Daphnia--A Handy Guide for the Classroom Teacher	138
The Use of Allelopathic Interactions as a Laboratory Exercise	142
Imbalance in Aquatic Ecosystems: A Simple Experimental Demonstration	145
A Hands-On Simulation of Natural Selection in an Imaginary Organism, *Platysoma apoda*	150
Environmental Pollution Effects Demonstrated by Metal Adsorption in Lichens	160
Using *Lemna* To Study Geometric Population Growth	164
Introducing Students to Population Genetics and the Hardy-Weinberg Principle	171

General Techniques

Preparing and Diluting Solutions: An Exercise for Courses in Biology Teaching Methods	182
Simple Principles of Data Analysis	184

Cell and Molecular Biology

Genetic Transformation of Bacteria

Robert Moss, Yeshiva University, New York, New York

The genetic makeup of an organism can be changed by mutating its DNA, or by inserting exogenous genes into that organism's DNA. Experiments showing that the addition of new genes can create a heritable change in an organism's phenotype provided the first conclusive evidence that DNA is the genetic material.

Natural transformation of bacteria was first observed in *Pneumococcus*. In 1944, strains of these bacteria were shown to be able to take up DNA fragments harboring beneficial genes from the surrounding medium. These results, which suggested that DNA was the genetic material, were not immediately accepted, as biologists of that day felt that only proteins contained enough complexity to carry genetic information.

Eight years later, other experiments confirmed the earlier findings. Scientists found that when a virus infected a bacterial cell, only the virus' DNA entered the cell and directed the formation of new viruses. The viral coat protein remained on the outer surface of the cell.

These experiments, along with other later DNA transfer experiments, definitively establish the role of DNA as the genetic material.

Molecular biologists can now alter the genetic makeup of many organisms by adding genes to their DNA. The entire field of modern molecular biology relies upon the ability of bacteria to take up and express foreign genes.

Many organisms have evolved the capability of adding exogenous DNA to their genomes; presumably this process evolved to supply the cell with new genetic information.

However, many of the cell types most useful in biological research, such as *E. coli* and human cells, do not have the ability to spontaneously take up exogenous DNA. Yet even these cells can be forced to take up and express exogenous genetic material under certain conditions in a procedure known as DNA-mediated cell transformation. Using this technique, bacteria may be "engineered" to mass produce nearly any DNA, RNA or protein molecule that biologists can design.

In this experiment, students will transform an ampicillin-sensitive strain of *E. coli* with a plasmid, pBR322, containing a gene for ampicillin resistance. After transforming the cells with the plasmid DNA, cells are plated on agar medium that has been supplemented with 50 µg/ml of ampicillin. Only the cells that have absorbed the DNA and express it will be able to survive the antibiotic treatment.

Procedures

Instruct the students on the proper sterile technique. Have them wipe down and sterilize their work area. All procedures must be carried out using the sterile technique. (See

NABT's publication, *Working with DNA & Bacteria in Precollege Science Classrooms* for standard microbiological practices and aseptic techniques.)

Genetic Transformation of Bacteria

Preparation of Competent Cells

Bacteria must be treated with calcium salts before they are able to take up exogenous DNA. These treated cells are referred to as 'competent.'

Grow a culture of an ampicillin-sensitive strain of *E. coli* (such as strain C600, B or HB101) in any rich bacterial medium (see Table 1) to mid-logarithmic phase (an absorbance value of approximately 0.6 at 600 nm).

Place the culture on ice and keep the cells chilled through the entire procedure.

Centrifuge 20 ml of the culture in a refrigerated clinical centrifuge at 5000 RPM for five minutes. If a refrigerated centrifuge is not available, simply chill the rotor buckets before use instead.

Pour off the supernatant, taking care not to contaminate the tube.

Use a sterile Pasteur pipet to resuspend the pellet in the few drops of supernatant remaining in the tube.

Add 10 ml of sterile, cold 0.05M $CaCl_2$ to the cells and mix well on a vortex mixer.

Incubate the cell suspension in the ice bath for 10 minutes.

Centrifuge the cell suspension a second time at 5000 RPM for five minutes and pour off the supernatant.

Resuspend the cells in the few drops of medium remaining on the pellet.

Add 1 ml sterile, cold 0.05M $CaCl_2$ to the cells and vortex again.

Place the cell suspension on ice for 10-20 minutes.

Table 1. Solutions.

LB Bacterial Broth:
10 g Bacto-tryptone
5 g yeast extract
10 g NaCl
950 ml H_2O
pH to 7.0 with 5N NaOH
Add water to 1 liter. Autoclave to sterilize.

Transformation Buffer:
10 mM $CaCl_2$
10 mM $MgCl_2$
10 mM Tris, pH 7.0
Autoclave to sterilize.

DNA Storage Buffer:
10 mM Tris, pH 7.5
1 mM EDTA
Autoclave to sterilize.

Table 2. Sample preparation.

TUBE #	DNA	BUFFER	CELLS
1	---------	.15 ml	0.15 ml
2	50 µl	.1 ml	0.15 ml
3	50 µl	.1 ml	0.15 ml + 1 µl 25 µg/ml DNAse

DNA is 1 µg/ml pBR322 in DNA storage buffer.
For tube 3, incubate DNA, buffer and DNAse at 37°C for 10 minutes BEFORE adding cells.

Genetic Transformation of Bacteria

Transformation of Competent *E. coli*

Obtain three sterile, capped tubes; number them 1 through 3. Add the reagents as described in Table 2. Keep tubes on ice until ready to incubate. Use sterile technique at each step.

Incubate the tubes on ice for 20-30 minutes; at 37° C for exactly two minutes; then remove the tubes and place at room temperature.

Add 1 ml of sterile medium without antibiotics to each tube, and incubate at 37° C for 40 minutes. The ampicillin-resistant gene on the plasmid is expressed during this period to confer resistance to transformed cells.

Obtain six plates with antibiotics and six without. Mark the six with antibiotics as follows:

#1 20 µl + Amp
#1 200 µl + Amp
#2 20 µl + Amp
#2 200 µl + Amp
#3 20 µl + Amp
#3 200 µl + Amp

Mark the six without antibiotics similarly, except using "___" instead of "+Amp." Pipet the appropriate amount of each numbered culture onto each plate. Two different volumes are used to maximize the possibility of getting a plate with a quantifiable density of transformed colonies. Spread the bacteria on the plates and incubate them at 37° C.

Table 3. Reagents.

REAGENT	COMPANY
pBR322	Sigma Chemical Company
Ampicillin	Phone: 800-325-3010
DNAse	
Yeast extract	
E. coli strain B	American Type Culture Collection
	Phone: 800-638-6597
Difco Bacto-Tryptone	Baxter Scientific Products
	Phone: 800-526-2193

Important Publisher's Note: Information on ordering reagents has been provided by the author for your ease. However, the National Association of Biology Teachers recognizes that there are other supply companies, and is in no way endorsing these firms or suggesting that they are the sole providers of these materials.

Students should return in 24-48 hours to check their plates. Estimate the number of colonies on the plate or note the density of the "lawn" if the plates are too dense to see individual colonies.

References

Sambrook, J., Fritsch, E.F. & Maniatis, T. (1989). *Molecular cloning: A laboratory manual* (2nd ed., vol. 3). Cold Spring Harbor, NY: Cold Spring Harbor Laboratory Press.

Alberts, B., Bray, D., Lewis, J., Raff, M., Roberts, K. & Watson, J.D. (1989). *Molecular biology of the cell* (2nd ed.). New York: Garland Publishing.

A Simple Demonstration of Fermentation

William J. Yurkiewicz, Millersville University, Millersville, Pennsylvania
David S. Ostrovsky, Millersville University, Millersville, Pennsylvania
Carole B. Knickerbocker, Pennsylvania Department of Environmental Resources, Conshonocken, Pennsylvania

In 1897, Eduard Buchner first demonstrated that yeast extracts could convert sugar into alcohol with a release of carbon dioxide. This observation was a critical starting point for modern research in metabolism and enzymology. Over the next half century, the analysis of fermentation showed that this metabolic pathway includes the reactions of glycolysis. Glycolysis, a universal mechanism for generating ATP in anaerobic organisms, serves as the starting point for ATP production in aerobic organisms.

Unlike a great many metabolic pathways, fermentation --and the various parameters that influence it-- can be easily studied with a very simple experimental procedure that has consistently given good results. Students could then spend most of their time on what we consider to be the activities of prime importance in the laboratory: the development of hypotheses that can easily be tested with our setup, the design of experiments to test these hypotheses, and the careful analyses of data. Even though the setup and procedure are fairly simple and straightforward, the design of specific experiments and data analyses can be quite sophisticated. The procedures described here have been tested in a number of undergraduate laboratory sections, as well as in an independent study project. They have been found to work well.

Materials

- Disposable 1 ml serological pipets
- Disposable critocaps
- Graduate cylinders (cylinders should be at least 23 cm in height)
- Thermometers
- A water bath or a large container for warm water that will hold the graduate cylinders
- 250 ml Erlenmeyer flasks, test tubes, pipets
- Dry active yeast
- Sucrose (table sugar)
- Buffer tablets

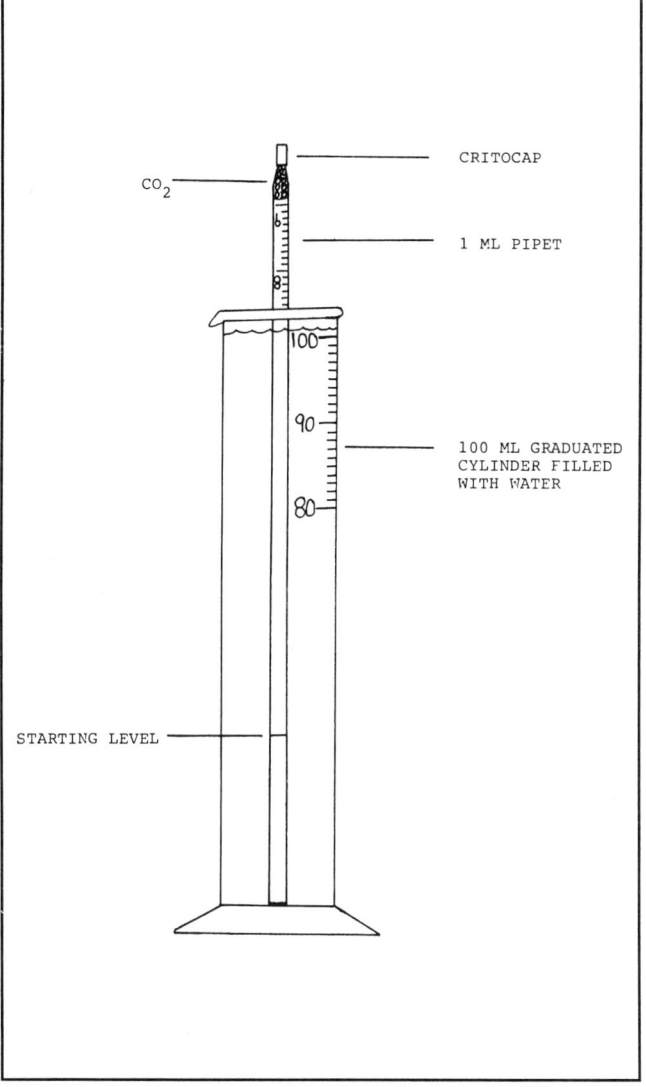

Figure 1. Simple apparatus for measuring carbon dioxide production during fermentation.

A Simple Demonstration of Fermentation

Procedure

1. Dissolve one pack (8 grams) of yeast in 100 ml of tepid tap water.

2. Make a 20% sucrose solution with tepid tap water.

3. Dissolve one buffer tablet (pH 3 is best) in 50 ml of tap water.

4. Into one test tube, pipet 1 ml of yeast solution, 1 ml of sucrose solution and 2 ml of buffer solution. This will yield .08 grams of yeast suspended in a 5% sucrose solution at pH 3.

5. Let the test tube stay in the water bath at 37 °C for 10 minutes. This will give the yeast time to take in the sucrose and begin fermentation.

6. Draw 1 ml of the yeast suspension into the 1 ml pipet. Cap the pointed end of the pipet with a critocap and place the pipet pointed end up in a graduate cylinder filled with water at 37° C. The cylinder should be kept in the water bath.

7. As carbon dioxide is released, the level of the yeast suspension in the pipet will descend. Record this level at five-minute intervals. The change in level divided by the elapsed time will give the rate of carbon dioxide production.

Projects

Using this basic setup, it is easy to vary parameters such as pH sucrose concentration, yeast concentration, temperature, type of carbohydrate used (i.e., sucrose, glucose, starch, etc.), and end product (ethanol) concentration. The simplicity of the setup (Figure 1) makes it ideal for general laboratories; its versatility challenges the students to formulate hypotheses concerning fermentation, to design experiments to test these hypotheses and to evaluate the data. The typical results are outlined in Table 1.

References

Angier, N. (1986). A stupid cell with all the answers. *Discover, 7*(11), 71.

Jagadish, M.N. & Carter, B.L.A. (1978). Effects of temperature and nutritional conditions on mitotic cell cycle of *Saccharomyces cerevisae*. *Journal of Cell Science, 31*, 71.

Vanoni, M., Vai, M. & Frascotti, G. (1984). Effects of temperature on yeast cell cycle analyzed by flow chemistry. *Cytometry, 5*, 530.

A Simple Demonstration of Fermentation

Table 1. The effect of pH on fermentation.

pH	2	3	4	6.4
Dextrose concentration	10%	10%	10%	10%
Yeast concentration	8%	8%	8%	8%
Temperature	24°C	24°C	24°C	24°C
Minutes	ml of CO_2 produced			
5	0	.057	.027	0
10	0	.280	.190	.106
15	.033	.580	.423	.247
20	.133	.793	.657	.477
25	.243	1.0	.780	.657
30	.380		.917	.843
35	.507		1.0	.973
40	.607			1.0
45	.680			
50	.770			
55	.847			
60	.893			
65	.983			
70	1.0			

Biology Labs That Work: The Best of How-To-Do-Its

Demonstrating the Effects of Stress on Cellular Membranes

Darrell S. Vodopich, Baylor University, Waco, Texas
Randy Moore, University of Akron, Akron, Ohio

Living beet cells are excellent models for some simple experiments involving cellular membranes. Membranes are functionally important because they separate and organize chemicals and reactions within cells by allowing selective passage of materials across their boundaries. As in all biology, a membrane's structure relates to its function, and an understanding of membrane function is fundamental for introductory biology students.

Unfortunately, most laboratory experiments investigating characteristics of membranes are too complex for introductory biology courses or include artificial rather than living membranes. This paper describes two simple procedures allowing students (grades 7-12) to experiment with living membranes and to relate their results to fundamental membrane structure.

The membranes of living eukaryotic cells, including beet cells, consist of a bilayer of phospholipid molecules interspersed with protein molecules. A phospholipid molecule is a combination of a phosphate group and two fatty acids bonded to a three-carbon

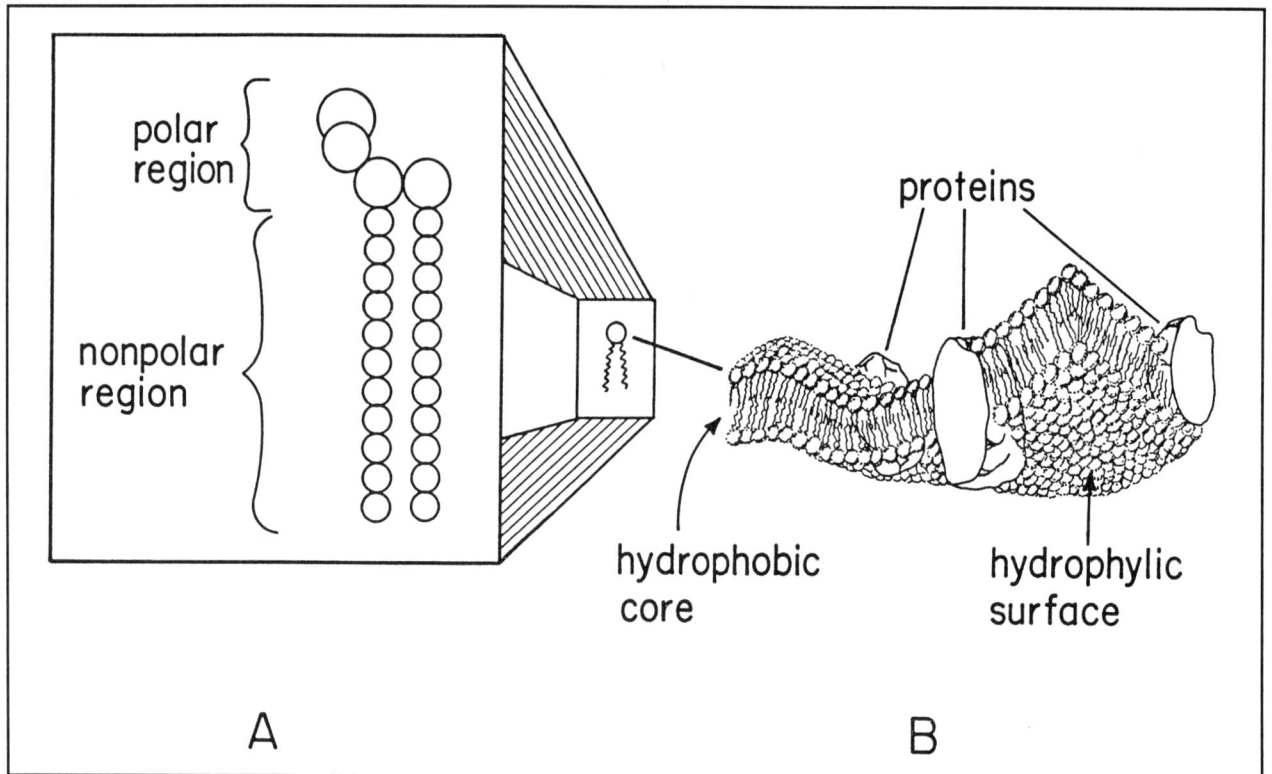

Figure 1(a) A phospholipid molecule. (b) Model of a bilayer membrane.

glycerol chain (Figure 1a). The resulting phospholipid molecule is polarized: The polar (charged) phosphate group is hydrophilic (water-loving) and the nonpolar fatty-acid groups are hydrophobic (water-fearing).

Effects of Stress on Cellular Membranes

Polarized phospholipids will innately self-assemble into a double-layered sheet of molecules forming a membrane. The hydrophobic tails of the lipids form the core of the membrane and the hydrophilic groups line both surfaces (Figure 1b). This elegant assembly is stable and allows selective penetration by small, lipid-soluble, hydrophobic molecules. The lipid bilayer resists penetration by most large, hydrophilic molecules.

Roots of beets (*Beta vulgaris*) contain an abundant red pigment called betacyanin, which is localized almost entirely in the large central vacuoles of the beet cells. These vacuoles are surrounded by a vacuolar membrane (i.e., tonoplast) and the entire beet cell is further surrounded by a cellular or plasma membrane. As long as the cells and their membranes are intact, the betacyanin will remain inside the vacuoles. However, if the membranes are stressed or damaged, betacyanin will leak through the tonoplast and plasma membrane and produce a red color in the water surrounding the stressed beet. This red color allows a student to easily assess damage to living membranes by monitoring the intensity of color produced by stressful, experimental treatments such as extreme temperatures or lipid-dissolving solvents.

The Effect of Temperature Stress on Membranes

Extreme temperatures provide a good set of treatments for student experimentation because high or low temperatures can physically destroy a membrane. In addition, temperatures can be easily and accurately measured. To prepare for such an experiment, you'll need:

- Fresh beets
- A freezer
- A beaker
- Six test tubes
- A thermometer
- A cork borer
- A metric ruler
- A test-tube rack
- A refrigerator
- Forceps
- A razor blade

Use the cork borer and razor blade to cut six sections of beet tissue into cylinders 15 mm long and 5 mm in diameter. Rinse the beet sections to remove pigment from the damaged cells. Each of the sections will be subjected to one of the temperatures listed in Table 1.

For the two coldest treatments, place two beet sections in two labeled test tubes and place one tube in a freezer (-5° C) and one tube in a refrigerator (5° C) for 30 minutes.

Table 1. The color intensity of betacyanin leaked from damaged membranes treated at six temperatures.

Tube Number	Treatment (°C)	Color Intensity (0-10)	Absorbance (460 nm)
1	70		
2	55		
3	40		
4	20	.	
5	5		
6	-5		

Biology Labs That Work: The Best of How-To-Do-Its

Effects of Stress on Cellular Membranes

Then add 10 ml of water to each test tube and place them in a rack at room temperature.

For the warmer treatments (i.e., 20, 40, 55, 70° C), heat a beaker of water to 70° C and submerge a beet section in the water for one minute. Place the section in a labeled test tube with 10 ml of water at room temperature. Cool the beaker of water using ice or tap water to 55° C and submerge another section for one minute. Place this section in a labeled test tube with 10 ml of water at room temperature. Repeat the procedure of cooling and submersion for each of the remaining temperature treatments. After completing all of the treatments you should have a rack of six labeled test tubes, each with 10 ml of water at room temperature and a beet section that has been subjected to a different temperature. Shake the tubes occasionally and allow 30 minutes for the pigment to leak out of the stressed cells. Then remove the beet sections from the tubes.

While students wait for the experiments to proceed, you might discuss the construction of graphs to display the results or consider the implications of all membranes having similar structure.

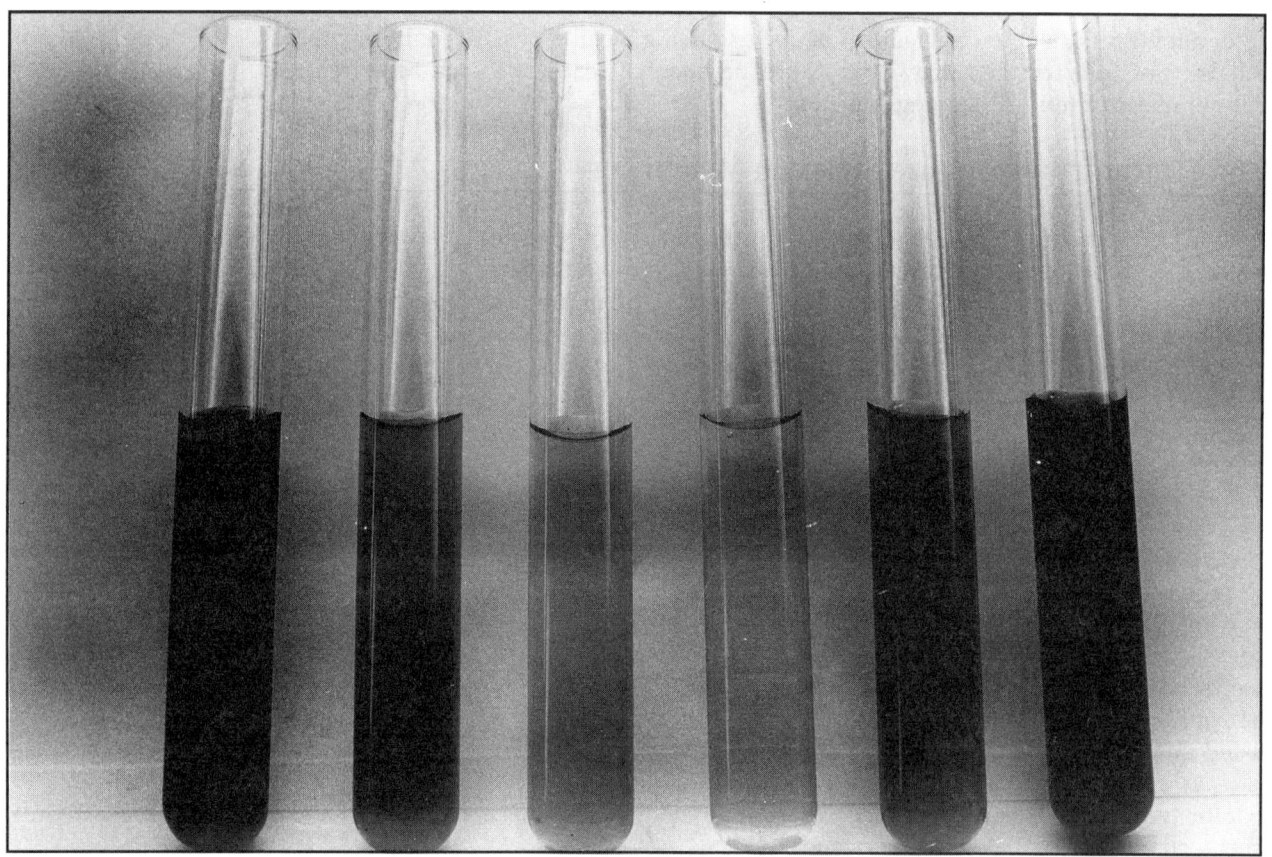

Figure 2. Six solutions of betacyanin from a beet tissue treated at six temperatures. From left to right, treatments were -5, 5, 20, 40, 55 and 70° C.

Effects of Stress on Cellular Membranes

The water surrounding the stressed beets will contain various amounts of betacyanin (Figure 2). You can assess the relative damage or stress caused by each temperature treatment by comparing the intensity of color in each tube. Although a spectrophotometer will provide the most accurate color readings, middle school instructors may wish to have the students estimate the color using a subjective scale. We suggest using values 1-10 as a subjective scale measuring color intensity. The darkest solution would have a value of 10 and the lightest a value of 1. Record your results in Table 1.

Questions for students

1. Which temperature stressed and damaged the membranes the most?
2. Exactly how could high temperatures tear a membrane?
3. Did low temperatures stress the membranes by the same mechanism as high temperatures?

Spectrophotometric Analysis

Students can use spectrophotometers to objectively assess the relative amounts of betacyanin resulting from membrane damage. Inexpensive spectrophotometers (colorimeters) can easily measure the absorbance of 460 nm light by betacyanin. This light absorbance is a direct measure of the concentration of betacyanin and an indirect measurement of membrane damage. Although you can assess the results of the above treatments without electronic equipment, use of a spectrophotometer enhances the experiments by quantifying the results.

After making and recording the absorbance readings for each temperature, these data can easily be plotted on X-Y axes. Plot temperature, the independent variable, on the X-axis and absorbance on the Y-axis.

Questions for students

1. Did any two treatments produce solutions of similar color intensity?
2. What is the advantage of using a machine rather than your eyesight to measure color intensity?

The Effect of Organic Solvent Stress on Membranes

The lipid structure of membranes can be altered by organic solvents that dissolve a membrane's lipid component. Acetone and alcohol are readily available solvents that severely stress membranes. We suggest an experiment that compares the membrane disruption (lipid solubility) of acetone with that of alcohol and tests the effects of various concentrations of each solvent. **Be sure to warn students of the hazards of organic**

Effects of Stress on Cellular Membranes

solvents. Most solvents, such as acetone, are flammable and volatile. Students must avoid breathing fumes and avoid any skin contact with the solvents.

To prepare for an experiment on the effects of solvents on membranes, you'll need:

- Fresh beets
- A metric ruler
- 10 ml of methanol or ethanol
- A cork borer
- A graduated cylinder
- Six test tubes
- A razor blade
- 10 ml of acetone
- A test-tube rack

Prepare 1%, 25%, and 50% (v/v) solutions of acetone in water, and three more solutions of the same concentrations using methanol in water. Cut and rinse six beet sections as described in the experiment on temperature stress (see page 15). Place one section in each of six labeled test tubes and add 10 ml of one of the six solvents to each test tube. Seal the test tubes with corks to prevent fumes from escaping. After 30 minutes, remove the beet sections and compare the red color of each solution. Record the color intensity of each tube in Table 2, using a subjective scale of 1-10 as described in the experiment on temperature stress.

Table 2. The color intensity of betacyanin leaked from damaged membranes treated with three concentrations of two organic solvents.

Tube Number	Treatment	Color Intensity (0-10)	Absorbance (460 nm)
1	1% Acetone		
2	25% Acetone		
3	50% Acetone		
4	1% Methanol		
5	25% Methanol		
6	50% Methanol		

Questions for students

1. Which solvent stressed the membranes more?
2. Did higher concentrations of the solvents cause more damage?
3. Are lipids soluble in both acetone and alcohol?
4. Which solvent would you conclude has the greatest lipid solubility?

Demonstrating Osmosis and Anthocyanins Using Purple Onion

Anne A. Kamrin, The Baldwin School, Bryn Mawr, Pennsylvania
Joan S. LaVan, The Baldwin School, Bryn Mawr, Pennsylvania

Wet mounts of white onion cells are widely used in introductory biology to demonstrate plant cell structure. We have found that purple onion cells show cellular structure more clearly and can also be used to directly observe osmotic changes in cells under a microscope rather than resorting to the use of models. These studies are simple to perform, require only ordinary equipment and use easily obtainable materials.

Students are asked to make tear preparations and wet mounts of the outer, purple epidermis of the purple onion bulb's leaf scales. The vacuoles of these cells contain an anthocyanin pigment that delineates the large, central plant cell vacuole and makes it possible to observe the positions of cell and vacuolar membranes that are usually very difficult to discern in white onion preparations. In addition, the position of the cell wall in relation to these membranes and to the nucleus are clear in the purple onion (Figure 1). Students may also be given a "feel" for the condition of plasmolyzed and non-plasmolyzed cells. Finally, the study of the purple onion epidermis is open-ended and can be used as a starting point for other projects such as investigation of vacuolar pigments as pH indicators.

Procedures

The tear preparations are mounted in distilled water as well as in NaCl solutions of varying osmotic concentration to demonstrate plasmolysis. Then, the students can remove the various mounting solutions by capillarity using paper towels and can re-flood the preparations with other solutions, thereby changing the environment of the cells. For example, a cell mounted in 2% NaCl will plasmolyze (Figure 2); when it is re-flooded with distilled water it will return to a more normal condition (Figure 3). Continuing to add distilled water will produce a slight bulging as a result of increased turgor pressure.

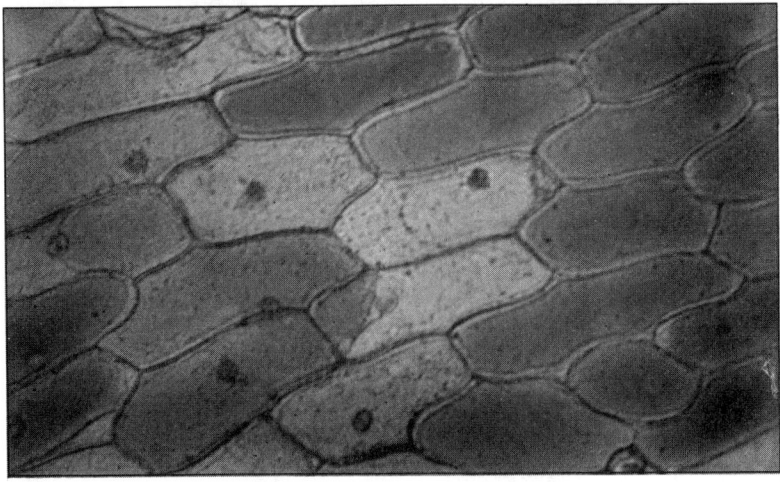

Figure 1. Normal condition of purple onion epidermal cells. Gray shaded areas show pigment-filled central vacuoles. Nuclei can be observed in some cells. 100✕.

Demonstrating Osmosis and Anthocyanins

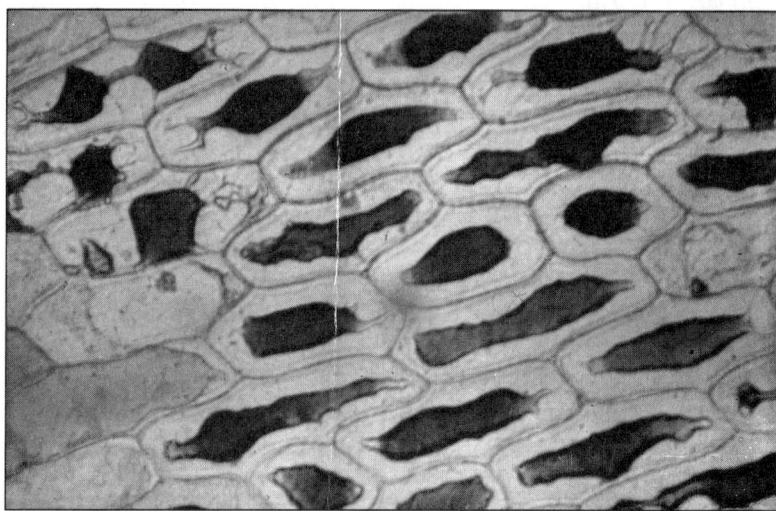

Figure 2. Condition of purple onion epidermal cells in 2% NaCl solution. Concentration of pigment in the central vacuoles of plasmolyzed cells can be seen. Some cytoplasmic strands are also visible. 100X.

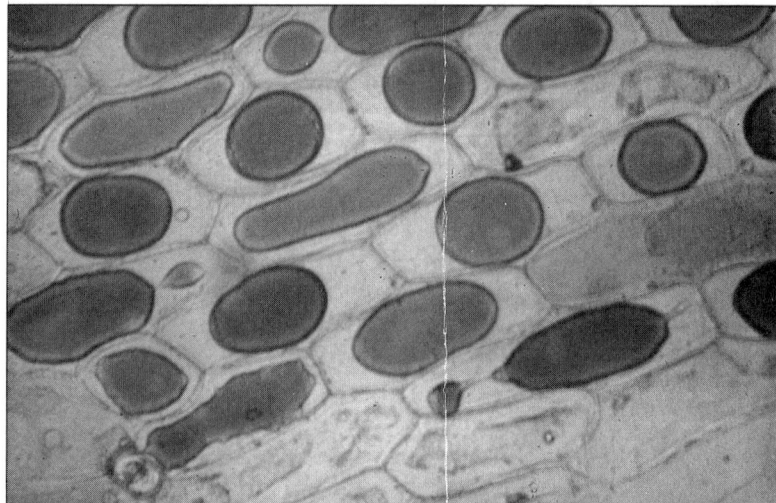

Figure 3. Condition of plasmolyzed purple onion epidermal cells rehydrating following flooding with distilled water. The vacuoles become rounded. Some areas of the cell walls show the position of the plasmodesmata. 100X.

Students can actually watch the changes occurring in cells within a short period of time. The idea that osmosis involves movement of water across membranes is emphasized because the anthocyanin pigment remains within cell vacuoles, and its color intensity changes due to changes in pigment concentration as water moves into or out of the cell, as is shown in Figures 1-3. The rate at which these changes occur can be observed. These observations readily lead to a discussion of the causes of wilting, turgor and the role of the cell wall in preventing the cytolysis of plant cells.

Studies of Anthocyanin

For more advanced students with some background in chemistry, further studies of anthocyanin pigments can be undertaken. Anthocyanin pigments are pH indicators. They generally appear to be red-purple in solutions around pH 2, while in solutions above pH 4.5 they appear blue-purple. Above neutrality, they tend to break down and appear green. These color changes can be demonstrated using the purple onion cells from the previous preparations. The addition of 1N acetic acid (pH 2.4) will cause vacuoles to take on a reddish hue, while addition of 0.1N boric acid (pH 5.2) will render the vacuoles bluish. The use of ammonia or strong bases to elicit the basic color change is to be avoided since it produces an irreversible change in the pigment, thus destroying its usefulness as an indicator.

Students should be informed that there are a number of different anthocyanin pigments that all have similar basic structures, with different active groups that produce variations in color. This can lead to a discussion of the role these pigments play in the coloration of flowers, fruits and leaves, as well as the effect that the external environment has on the colors produced.

Reference

Clevenger, S. (1964). Flower pigments. *Scientific American, 210*(6): 84-92.

pH and Rate of Enzymatic Reactions

Roy B. Clariana, EG&G Rocky Flats Plant, Golden, Colorado

This article describes a quantitative and very inexpensive way to measure the rate of an enzymatic reaction. The laboratory activity deals with the effects of different pH levels on the rate of reaction; however, the methodology can be easily adapted for measuring the effects of temperature, substrate concentration, enzyme concentration and reaction inhibitors. Also, the approach can be used either as a demonstration or as a student laboratory activity.

Molecules of hydrogen peroxide are broken down in the presence of the enzyme catalase. A yeast suspension is used to provide the catalase, although we have also used crushed potatoes with the same results. Small filter paper squares are dipped into a yeast suspension and then dropped in a very dilute hydrogen peroxide solution. The filter paper square sinks to the bottom due to its density. (Note: Filter paper is used because ordinary paper will not easily sink when dropped in water.) The catalase on the filter paper begins to react with the hydrogen peroxide, producing small oxygen bubbles. These bubbles are trapped in the fibers of the filter paper square. Eventually the paper will float to the surface (Figure 1). The time required for this to occur depends directly on the rate of the reaction. Obviously, the faster the reaction, the faster the paper will float. The optimum pH for catalase function is 6.8. The further the pH is from the optimum value, the slower the reaction rate. Thus in this activity, a graph of the rate of reaction versus pH will produce a normal curve with an optimum at pH 7.

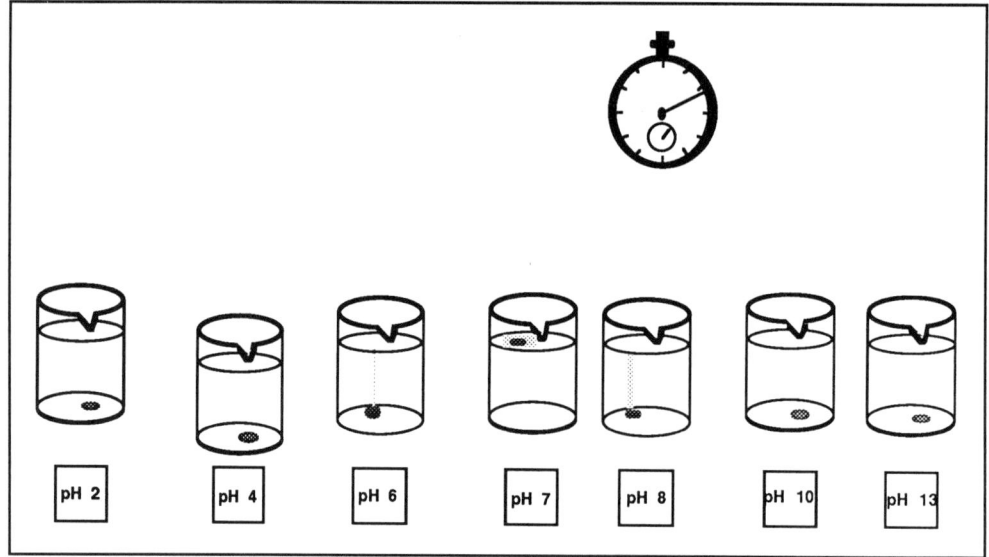

Figure 1. These beakers contain dilute hydrogen peroxide solutions. The enzyme on the square pieces of filter paper in each cylinder causes the production of oxygen which then causes the pieces to float. The rate of enzymatic reaction was highest at pH 7, as indicated by the floating piece of paper in that cylinder.

Materials

- Hydrogen peroxide
- Packet of yeast
- Scissors (or razor blade)
- Ruler (mm)
- Stopwatch
- Several small forceps
- Filter paper (e.g., number 3)
- Seven beakers for each group of students (approx. 500 ml)
- Several containers for preparing solutions
- Paper labels and pen
- Hydrochloric acid and sodium hydroxide

Preparation

Prepare the following materials beforehand (or have the students prepare the following):

1. Mix the yeast in a beaker containing warm water. Exact measurements of water and yeast are not critical. Allow about 30 minutes for the yeast to activate. Label this mixture "catalase" or "enzyme."

2. Cut the number 3 filter paper circles into squares that are 5 mm on a side. Try to avoid excessive contact with the paper and be sure your hands are as clean as possible. Oils from your hands can alter the paper's ability to absorb the catalase solution.

3. Using water (tap water is OK), prepare a 1:1000 dilution of the hydrogen peroxide in the 500 ml cylinder. Inexpensive and easily obtained hydrogen peroxide from a supermarket or drug store works fine. Depending upon the freshness of this solution, additional dilution may be required. To test this solution, pour about 50 ml of the diluted hydrogen peroxide into a beaker. Use the forceps to dip a filter paper square into the catalase (i.e., yeast) mixture and drop the filter paper square into the beaker. The paper should sink completely to the bottom and then float to the top in about 40 seconds. If it takes less time, add water to further dilute the peroxide; if it takes too much time, add more hydrogen peroxide and test again. You will probably have to add water—a little hydrogen peroxide goes a long way.

4. Fill seven beakers each about two-thirds full with the hydrogen peroxide solution from Step 3 and label these "pH 2, pH 4, pH 6, pH 7, pH 8, pH 10 and pH 12." Using hydrochloric acid and sodium hydroxide, adjust each to the desired pH value (Lennox & Kuchera 1986). Adding acid and base in this step will slightly lower the concentration of the hydroxide peroxide solutions. To correct this, add an approximately equal volume of tap water to each to maintain the dilution concentration.

pH and Rate of Enzymatic Reactions

Procedure

1. Use forceps to dip a filter paper square into the catalase solution.
2. Drop the square into the first beaker. It will sink to the bottom.
3. Start the stopwatch immediately. Record the time required for the filter paper square to float to the surface (i.e., horizontal). Record this time data.
4. Repeat twice more for the first beaker (to obtain an average for that beaker).
5. Repeat steps 1 through 4 for the remaining beakers.
6. Have the students graph the results with pH as the horizontal axis and rate (i.e., 1/t) as the vertical axis. Rate is determined by taking the inverse of time (e.g., 40 seconds becomes 0.025).

Results & Discussion

The graph of this demonstration resembles a normal distribution (i.e., bell-shaped curve). Like all enzymes, catalase must be in a specific form or shape for the enzymatic reaction to progress. This specific shape alters at different pH levels, thus rendering the enzyme less effective. The optimal pH of an enzyme depends upon its location within living organisms. For example, certain digestive enzymes work best in acidic environments. In life there is always a remarkable matching of function and form even at the molecular level. This demonstration may help your students to grasp this fundamental concept of life as well as observe experimentally the results of the mechanism of enzymatic reactions.

Reference

Lennox, J.E. & Kuchera, M.J. (1986). pH and microbial growth. *The American Biology Teacher*, 48(4), 239-241.

A Simple DNA Isolation Technique Using Halophilic Bacteria

Patrick Guilfoile, Whitehead Institute for Biomedical Research, Cambridge, Massachusetts

Biotechnology is an increasingly important topic in biology courses. Yet it is often difficult to present biotechnology concepts because of the lack of simple, inexpensive lab exercises.

One example of an important biotechnology procedure is the isolation of DNA. Once isolated, the DNA can be used in a variety of ways to genetically alter organisms (Watson, Tooze & Kurtz 1983). Existing techniques for DNA isolation are either expensive or involve procedures and chemicals that are difficult to use in a typical high school biology classroom (Myers 1985).

This article describes a simple technique for isolating DNA from halophilic bacteria. Compared to other protocols, this procedure is less expensive, faster (it requires only one class period) and does not require the use of potentially dangerous chemicals such as sodium perchlorate and chloroform.

Halophilic bacteria are an unusual group of microorganisms that grow only in environments with a high salt concentration (4-5M NaCl). Research on this salty lifestyle has focused on the cell envelope (Steensland & Larsen 1969; Mescher, Strominger & Watson 1974). Work by Steensland and Larsen (1969) indicates that the cell wall of these organisms disintegrates when put in a low-salt environment (below about 2 M NaCl).

The lysis of halophilic bacteria at low salt concentrations makes them ideally suited for use in a simplified DNA isolation lab by allowing the elimination of several steps normally required to disrupt the cells.

Materials

- Yeast extract
- Tryptone (Difco)
- Sodium chloride (reagent grade is best, but non-iodized table salt should work)
- Culture of *Halobacterium salinarum*
- One sterile test tube with 10 ml of liquid media
- A small centrifuge
- 1 ml distilled water or tap water
- 2 ml ice-cold 95% ethanol
- An eyedropper
- An inoculating loop or glass rod

Procedure (for 20 students)

About one week before the lab exercise:

Step 1. Prepare media [adapted from Steensland & Larsen (1969) and Mescher, Strominger & Watson (1974)].

DNA Isolation Technique

In one autoclavable container mix the following:

1 g tryptone
2 g yeast extract

Add water to make 100 ml of media.

In a second autoclavable container mix the following:

50 g NaCl
2 g $MgSO_4$
1 g KCl
.04 g $CaCl_2$

Add water to make 100 ml of media.

Autoclave separately, mix contents of the containers when cool. Aseptically dispense 10 ml aliquots into sterile test tubes.

Step 2. Inoculate test tubes with *Halobacterium salinarum*.

Step 3. Incubate at 30-37° C until the media in the tubes becomes turbid (two to seven days, depending on the size of the inoculum and the conditions of incubation).

In-Class Lab

Step 1. Centrifuge test tube at a speed high enough to pellet cells (approximately 3000 × g for five minutes).

Step 2. Pour off supernatant.

Step 3. Add 1 ml distilled water. (If you are working with very large numbers of bacteria, you should add 2 ml of distilled water and double the amount of ethanol used in Step 4.) Gently swirl or stir the contents of the test tube to ensure suspension and lysis of all cells. The liquid in the tube should become viscous.

Step 4. Carefully layer 2 ml of ice-cold 95% ethanol on top of the water layer. This can be done by slowly releasing it down the side of the test tube with an eyedropper. There should be a sharp interface between the water and ethanol layers. The DNA will precipitate at the water/ethanol boundary.

Step 5. With an inoculating loop or thin glass rod, begin twirling slowly at the interface. DNA should adhere to the rod.

The DNA thus isolated will be somewhat contaminated with cellular proteins and RNA.

Spectroscopic measurements demonstrate that the isolated material is, however, primarily DNA (Anonymous 1985). If additional purification is desired, any commonly employed deproteination procedure (e.g., Marmur 1961) can be used.

Additional Experiments

The original cell lysate or the isolated DNA can be used in viscosity experiments as described by Holt and Choe (1968).

Acknowledgments

I would like to thank my student, Eric Saletri, for his help in refining this exercise. Additional thanks to my fourth and seventh hour biology classes who tested the procedures described in this paper.

References

Anonymous (1985). *Bacterial DNA: Extraction and physical properties concept study kit.* Rochester, NY: Ward's Natural Science Establishment.

Holt, C.E. & Choe, D.T. (1968). Some experiments on the viscosity of bacterial DNA solutions. *The American Biology Teacher, 30,* 504-516.

Marmur, J. (1961). A procedure for the isolation of deoxyribonucleic acid from microorganisms. *Journal of Molecular Biology, 3,* 208-218.

Mescher, M.F., Strominger, J.L. & Watson, S.W. (1974). Protein and carbohydrate composition of the cell envelope of *Halobacterium salinarum. Journal of Bacteriology, 120,* 945-954.

Myers, R. (1985). A method for isolating DNA from *E. coli. The American Biology Teacher, 47,* 362-364.

Steensland, H. & Larsen, H. (1969). A study of the cell envelope of Halobacteria. *Journal of General Microbiology, 55,* 325-336.

Watson, J.D., Tooze, J. & Kurtz, D.T. (1983). *Recombinant DNA: A short course.* New York: W.H. Freeman Co.

A Paper Model of DNA Structure & Replication

Linda A. Sigismondi, University of Rio Grande, Rio Grande, Ohio

I have found that students have trouble visualizing the structure and replication of DNA. I have developed the following model (Figure 1) that I use concurrently with lecture and blackboard illustration to give individual students hands-on instruction.

Materials

- Two copies of Figure 1 on white paper
- Two copies of Figure 1 on colored paper
- Scissors

Procedure (DNA Structure)

1. Students cut out the pieces shown in Figure 1, placing the white pieces in one pile and colored pieces in another. (Alternately, I have prepared kits from posterboard prior to class. The kits save class time and can be used many times. The advantage in the paper model is that students can take it home and use it for studying.)

2. Students are told that each piece represents a nucleotide. Draw the structure of a

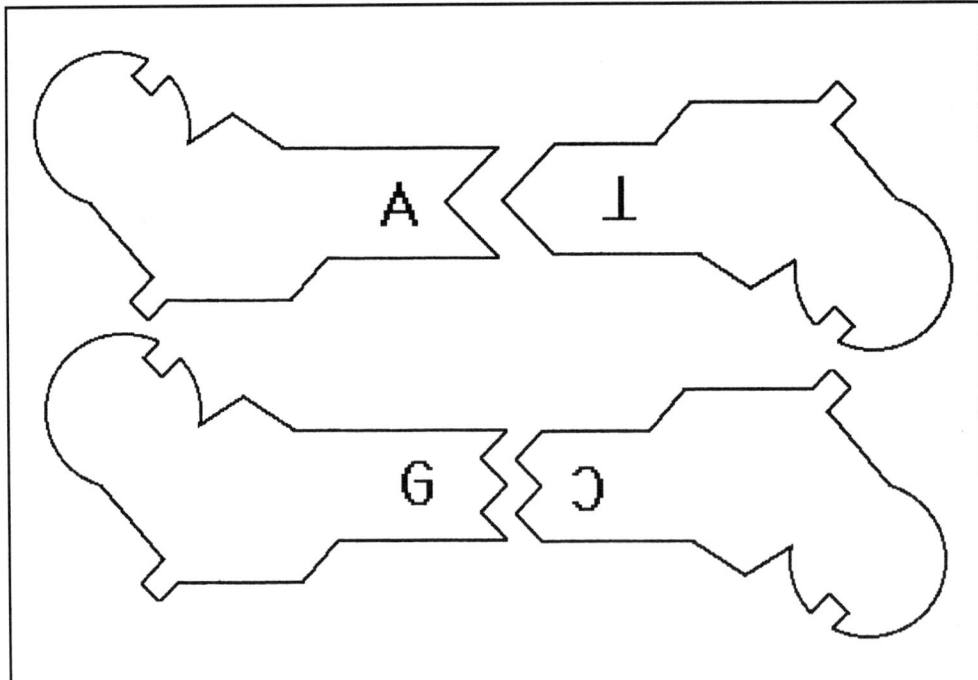

Figure 1. Models of nucleotides.

nucleotide on the blackboard and relate it to the paper structure by drawing a circle around the phosphate group, a pentagon around the sugar deoxyribose group and a rectangle around the base. Have students note that there are four different bases (A, T, C and G), each with a specific shape (corresponding to a specific chemical structure) and thus four different nucleotides. (For classes where detailed chemical structure is important, an instructor can draw the structural formula on the model prior to duplication or instruct the students to do so during the lecture.)

A Paper Model of DNA

3. Explain that the nucleotides in a DNA molecule are connected in a chain by chemical bonds between the sugar of one nucleotide and the phosphate of the one below it. Have students link together four of the white nucleotides by matching the tabs.

4. Next, explain that the DNA molecule consists of two such chains running in opposite directions with attractive forces between the bases holding them together. Have students note that due to the shape of the bases, only certain ones will fit together: A links only to T and G only to C. Students then find the matching white nucleotides to their original four and complete the structure of DNA, keeping in mind that the letters must be facing up so the resulting chains run in opposite directions.

   For advanced students, mention that the A and T bond by attraction at two points and that the G and C have three points of attachment. The bases G and A are purines composed of two rings while the bases C and T are pyrimidines consisting of only one ring. Thus the bases in G and A are longer than those in C and T.

5. Finally show an actual model of DNA illustrating the helical configuration of the molecule. If a molecular model is not available, I use the paper clip and pipe cleaner model by Peebles and Leonard (1987).

I sometimes relate this exercise to the original work on DNA structure by Watson and Crick (1953) who also used paper models as well as experimental data from other investigators to determine DNA structure. They dealt with the following information:

   a. There were always equal numbers of A and T in a molecule and equal numbers of G and C, but the numbers of A and T did not equal the numbers of G and C.

   b. The total number of purines (guanine and adenine) equaled the total number of pyrimidines (cytosine and thymine).

   c. The purine bases were longer than the pyrimidine bases.

   d. X-ray diffraction patterns indicated two chains coiled in a helix.

Have the students observe that their paper model is consistent with the above information.

A Paper Model of DNA

Procedure (DNA Replication)

1. Explain that the bonds between the bases are much weaker than the bonds forming the chains (hydrogen bonds versus covalent bonds), thus the two chains can come apart. Have students separate their two chains leaving some space between them.

2. Explain that there are additional nucleotides inside the nucleus of cells capable of attaching to the open chain. Have them match the colored nucleotides to both chains of white nucleotides, thus building two new DNA molecules. Tell them the colored nucleotides just represent ones that were not previously part of the DNA and not structurally or functionally different from the white nucleotides.

3. Have students observe that the two chains are identical (in structure and sequence of nucleotides) to each other and to the original. Also have them note that each of the two molecules consists partially of the original and partially of the new nucleotides, thus illustrating semi-conservative replication.

I have found that this hands-on approach greatly facilitates learning and leads to a better understanding of DNA structure and replication. I also have been able to adapt this model for several different levels of biology by using different amounts of detail.

This model is crude and can give only a two-dimensional picture of DNA structure. Models on this topic are available from biological supply companies. However, this paper model is a economical alternative that can be easily made and replaced.

References

Peebles, P. & Leonard, W.H. (1987). A hands-on approach to teaching about DNA structure and function. *The American Biology Teacher, 49*, 436-438.

Watson, J.D. & Crick, F.H. (1953). Genetical implications of the structure of deoxyribonucleic acid. *Nature, 171*, 964-967.

Microbes and Fungi

Pasteurized Milk as an Ecological System for Bacteria

Alan L. Gillen, Tomball High School/Tomball College, Tomball, Texas
Robert P. Williams, Baylor College of Medicine, Texas Medical Center, Waco, Texas

Science Processes in the Laboratory

Experiences in science courses offer students an opportunity to be active learners. They permit students to engage in activities that go beyond the memorization and regurgitation of facts.

Laboratory exercises, especially, give students a chance to participate in the scientific enterprise and to act like a scientist. This allows them to practice the skills of a scientist, often referred to as science process skills (Collette & Chiappetta 1994). These include observing, inferring, measuring, experimenting, collecting data, graphing and interpreting data (Funk et al 1979). These skills generally stimulate a great deal of thought, involvement and interest in science. We observed this interest during a laboratory on milk ecology designed for students in high school biology courses.

Bacteria are important organisms for man to study because they are found everywhere. The ubiquity of bacteria is illustrated by the fact that they are found from hot springs to polar regions and from ocean depths to mountaintops. Since milk is an ideal medium for growth of bacteria, it can be used to study microorganisms. Laboratory exercises developed to study bacteria can both improve students' knowledge of microorganisms essential to our well-being and improve students' inquiry skills.

Milk as a Medium for Bacteria

Bacteriologists have long been interested in milk and dairy products. Joseph Lister in his paper, "On the Lactic Fermentation, and its Bearings on Pathology," described his new discovery that a specific bacterium causes the "natural souring" of milk. Lord Lister, a devout Quaker physician, is well known for his part in developing antiseptic surgical methods and making contributions to the germ theory of disease. Through a series of creative experiments, Lister demonstrated that *Streptococcus lactis* was not the result of milk souring, but a cause of milk spoilage due to its fermentation and putrefaction. Lister thought of milk spoilage as a type of infectious "disease." Lister made the connection between how a specific bacterium causes "lactic ferment" in milk and how bacteria contaminated in surgical wounds often cause gangrene and pyemia in humans. Lister's laboratory studies of milk souring, in part, led to our modern understanding of pathology.

There are few foods so important to man as milk. Milk is an essential form of nutrition in which bacteria have an important role. Most bacteria are killed in milk through pasteurization; however, some bacteria remain (Tortora et al 1992). Bacteria are used in many processes in the dairy industry. Products such as buttermilk, yogurt, sour creams and some cheeses are usually made commercially by adding bacteria to milk

Pasteurized Milk as an Ecological System

(Alcamo 1994). The changes of milk into other products are types of milk spoilage.

Milk spoilage under proper conditions can lead to the formation of cheese. The more common instance of milk spoilage often takes place in the kitchen refrigerator or in the dairy case at the supermarket. Bacteria multiply slowly and ferment lactose in milk into acid, thereby "spoiling" the milk (Alcamo 1994).

In this investigation you will see how pasteurized milk acts as a growth medium for bacteria and provides a wonderful ecosystem for scientific inquiry.

Upon completion of the laboratory unit, the student will be able to:

1. *Measure* the pH of milk.
2. *Observe* and *describe* visible changes in milk.
3. *Grow* bacteria on nutrient agar plates.
4. *Identify* bacteria (by shape) growing in milk for 10 days.

Materials

- Whole pasteurized milk
- Skim milk
- Chocolate milk
- Buttermilk
- Nutrient agar
- pH paper, or pH probe
- Compound microscope
- Crystal violet
- Cotton swabs, or inoculating needles
- Graph paper, data paper
- Petri plates
- 250 ml beakers
- 250 ml Erlenmeyer flasks
- Microslides
- Gram stain kit
- Rogosa SL Agar (optional)
- Sabouraud Agar (optional)
- Trypticase Soy Agar (optional)

Procedure

1. Place four students in a group. Assign each group a particular type of milk and temperature for bacterial growth listed below:

   Group I: Whole milk, room temperature (25° C).
   Group II: Whole milk, refrigerator temperature (4° C).
   Group III: Whole milk, incubator temperature (37° C).
   Group IV: Whole milk boiled (100° C), then cooled to room temperature (25° C).
   Group V: Skim milk, room temperature (25° C).
   Group VI: Buttermilk, room temperature (25° C).
   Group VII: Chocolate milk, room temperature (25° C).

2. Place 125 mL of each type of milk assigned to the group in an Erlenmeyer flask.

Pasteurized Milk as an Ecological System

3. Record daily observations of the milk with respect to date, temperature, pH, odor, color and growth of bacteria on agar plates on laboratory data sheets. As observations are made, students should pay particular attention to physical changes in the milk's composition and record any odors, particles, "growths" or new liquids forming in the milk.

4. Using a pH probe (Figure 1) or pH paper, record the pH of milk daily. When using the pH probe, rinse the probe in distilled water between readings; contamination will be so minimal as to not affect results. At the end of the experiment, graph the pH record, with pH on the Y-axis and time (days) on the X-axis. The experiment should be conducted over 10 to 14 days to see the completion of changes in the milk.

5. Record the temperature using a Celsius thermometer.

6. Use cotton swabs or inoculating loops to transfer milk (some with bacteria) to petri dishes containing agar to grow bacteria. Count the number of colonies growing on the petri dish. This number will give a relative value of the amount of bacterial growth and is not meant to be quantitative.

7. Use a simple crystal violet stain or Gram stain on the bacteria to help identify shapes of bacteria. Identify the bacteria as a coccus, bacillus or spirillum using a compound microscope with an oil immersion lens, if available. To perform a simple stain, a student places a small amount of bacteria in a droplet of water on a glass slide and dries the slide in air. Next, the slide is passed briefly through a flame in a process called heat fixing. The slide then is flooded with crystal violet or methylene blue for about two minutes, washed with water and blotted dry. A technique for advanced high school students is the Gram stain. It differentiates bacteria into two groups—Gram-positive and Gram-negative (Alcamo 1994). The procedure is described in most general microbiology textbooks (i.e., Alcamo 1994 or Nester 1978).

8. After the experiment is completed, have students record their observations on

Figure 1. Students demonstrating pH probe in milk.

a data table and graph the pH of milk through time (Figure 2). Observations should include the day and date of experiment, descriptions of milk changes, temperature, pH, odor of the milk, description of bacterial colonies grown on agar in petri plates, and bacterial shape. A sample of such data is shown in Table 1 (see page 36).

Pasteurized Milk as an Ecological System

9. There are many variations to this experimental protocol, such as: varying the time of the experiment, changing incubation temperatures, using different milk types and growing microbes on various media. If available, use Trypticase Soy Agar to obtain a "lawn" of *Streptococcus*, *Bacillus* (with endospores) and other bacteria, such as *Campylobacter*. Rogosa SL agar seems to enhance the growth of *Lactobacillus* the best. Use Sabouraud Agar for culturing molds and other fungi.

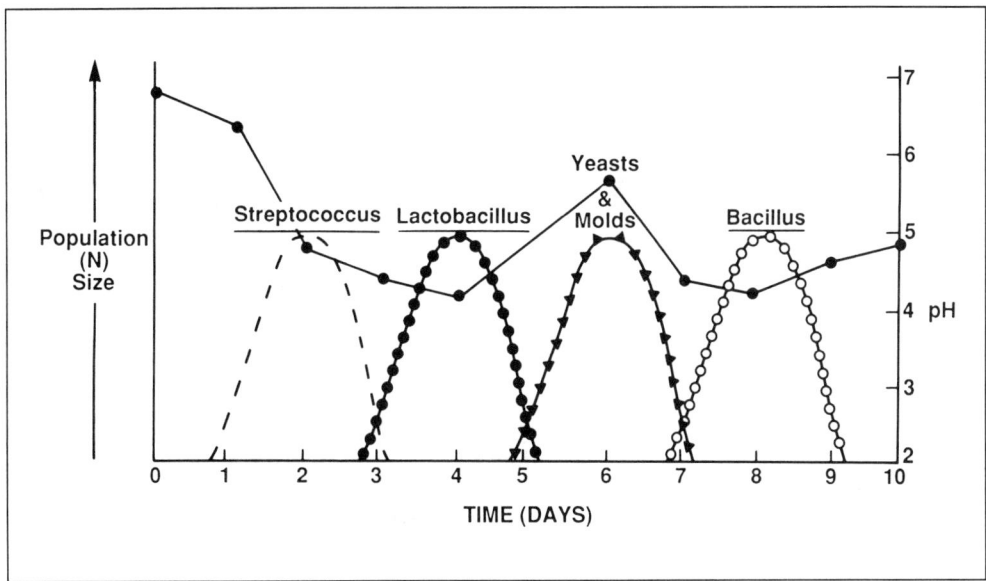

Figure 2. Ecological succession of bacteria and fungi in milk. General predominance (N) of microorganisms (dashed lines) through time due to environmental changes, as measured by pH (solid line).

Questioning

A wide range of questions should be asked students in this laboratory. The teacher should begin asking closed questions that recall the facts, followed by an analysis that will require students to tie facts together. Then, the teacher can proceed to ask open questions in order to direct the discussion. Finally, the teacher should question the students on the science processes they had to go through to successfully complete the laboratory. The student should get a feel for what a scientist may do in research.

Pasteurized Milk as an Ecological System

Table 1. Data sheet for changes in milk.

Milk Specimen: Whole pasteurized, room temperature   *Name: Jane Doe*

Day	Observations Noted	Temp. (°C)	pH	Odor	Color	Bacteria Colonies, Shapes and Gram Stain
0	normal milk color	11	6.7	fresh	solid white	• none
1	little change	24	6.4	slight	white	• in all quadrants, rods, Gram-positive
2	separation; solids and liquid	23	4.8	mild	variable white	• in all quadrants, cocci and rods, Gram-positive
3	coagulation	23	4.4	mild	white and dirty	• in all quadrants, cocci, Gram-positive
4	heavy coagulation	22	4.2	mild	solid white	• small colonies in all quadrants, cocci and rods, Gram-negative
5	like cotttage cheese	24	5.1	sour	creamy white with yellow liquid	• no observation
6	very chunky, like cottage cheese	23	5.7	mild	solid white with dirty liquid	• in 3 quadrants, mixed size, cocci and rods, Gram-negative
7	chunky, white, separation	24	4.4	mild	bright white	• 1 quadrant filled, cocci, Gram-positive
8	chunky, white	24	4.3	mild	bright white	• in all quadrants, rods, Gram-negative and cocci and rods, Gram-positive
9	chunky, white	24	4.9	mild	creamy white	• in all quadrants, cocci, Gram-positive
10	chunky, white	24	4.9	mild	creamy white	• in all quadrants, cocci, bacilli with spores, Gram-positive, many yeast cells

Examples of questions the teacher may want to use are:

1. What is pasteurization? Does pasteurization kill all the bacteria in milk? If not, what bacteria remain?

2. List three types of bacteria based on shape. Would you expect to find any of these in milk? Explain.

3. What is coagulation? At what time does it begin to occur at room temperature as milk changes? At what time does it occur at incubator and refrigerator temperatures?

4. What are four conditions for bacterial growth? Are these requirements met in milk experiments? Of these requirements for bacterial growth, which is naturally changed when you place milk in the refrigerator or incubator?

5. How do the results observed at room temperature compare with the results when milk is kept in the refrigerator? What advantage is there to keeping milk or other foods under refrigeration? Why?

6. Which milk changes the quickest? The slowest?

7. Of what benefit to the food industry is changed milk?

8. How long does it take for milk to change at room temperature? At refrigerated temperature?

9. Examine your pH graph. What is the relationship of milk pH to time (days)?

10. Look at your data and draw conclusions with respect to relationships between pH, temperature, types of milk and time. For example, compare type of milk (whole vs. skim milk) or temperature ranges (refrigerated vs. room temperature).

Discussion

The post-laboratory discussion is an excellent opportunity to solidify students' knowledge of the bacteria they studied. Go over the questions shown above that were prepared to guide students' thinking in the laboratory. In the postlaboratory, questions, discussion and conclusions should be drawn. The student should conclude that there is design and order in nature from the observation of the ecological succession of microbes. The ecological succession of microorganisms in unpasteurized milk (Figure 2) was discussed by Nester et al (1978). The succession of microbes in pasteurized milk follows a similar sequence.

In milk, the changing conditions bring about an ordered and predictable succession of microorganisms: first, streptococci; then a second bacterium, lactobacilli; then a third group, yeasts and molds; and finally *Bacillus* species. This sequence of changes is due to a changing chemical environment produced by the metabolic processes of the microorganisms.

Pasteurized Milk as an Ecological System

Streptococci break down the milk sugar, lactose, to lactic acid. The bacteria produce so much acid that they eventually inhibit their own growth and make the milk ideal for lactobacilli to grow. Lactobacilli multiply in this acidic environment and metabolize the rest of the lactose into more lactic acid until their growth is also inhibited by too much acid. Lactic acid sours the milk and curdles protein. Yeasts and molds grow well in this acidic environment and metabolize acid into nonacidic products. Finally, *Bacillus* (with endospores) species multiply in the environment where the only nutrient available is protein. *Bacillus* species metabolize protein into ammonia products and the pH rises (Figure 2). Also, *Bacillus* species excrete proteolytic enzymes that digest the remaining protein in the milk (Nester et al 1978). The odor of milk spoilage becomes apparent once this change has occurred.

In addition to microbes listed in Figure 2, sometimes we find other bacteria and fungi in unpasteurized milk, or milk that was poorly pasteurized. *Micrococcus* and *Proteus* attack the casein of milk. *Clostridium, Serratia marcescens, Achromobacter, Enterobacter aerogenes* (with capsules), and *Escherichia coli* have been found in milk that has been sitting around the lab. Finally, some students have observed *Campylobacter* in milk brought from a local dairy farm. We always exercise caution, using sterile technique, when working with and disposing of potential pathogens. Instructors might consult a local microbiology laboratory in their area for positive identification and proper disposal of these unusual bacteria found in milk.

Thus milk pH goes through a succession of changes with time, first due to fermentation and then, to putrefaction. These changes are brought about by microorganisms that have undergone a succession in their dominant population.

Another conclusion that could be drawn by the student is that cooling (refrigeration) retards bacterial growth. This fact is evident from data tables and/or pH graphs. The milk at room temperature spoils in a matter of one to two days, whereas the milk kept at refrigerated temperature may keep for four to ten days before changing.

Finally, we often use this laboratory as a starting point to discuss modern developments in the dairy industry and biotechnology. Agricultural applications of dairy bioengineering are discussed, including: 1) bovine somatotropin (BST) to increase output of milk in cows, and 2) enzyme chymosin, a substitute for renin used in making cheese. BST has generated some controversy because there are many local dairy farms near our school. This genetically engineered hormone increases the milk outputs of local cows and offers hope for greater profits by area farmers. BST is different from other bioengineered food products bioengineered in that it enhances the milk production and does not change the milk makeup. The genetically engineered product increases milk output by supplementing a cow's natural BST, a growth hormone produced by the pituitary gland. Milk from treated cows has been found to have the same nutritional value and composition as milk from untreated cows. Another production drug, used on a smaller scale, is the bioengineered enzyme chymosin, a substitute for renin used in making cheese. Sometimes, we add rennet liquid drops to milk in order to make cheese in the laboratory. This activity provides a natural extension to the milk laboratory and a visible application of dairy science.

This laboratory is one of several activities in a microbiology curriculum unit called *The Unseen World*, named after Dubos' book (1962) with the same title. *The Unseen World* has been test piloted in a Houston inner-city school, E.E. Worthington High School, and continues to be used at Tomball High School and Tomball College in Texas. The unit is designed to make the unseen world of microbes relevant to the average high school student. Many of the activities also can be adapted to middle schools. The laboratory utilizes materials that should be readily available at most middle or high schools. Materials and supplies required for this activity are relatively inexpensive. All activities in the unit are designed to challenge the students to think of the unseen as relevant to their own world.

Acknowledgments

This work was supported by the Houston Mathematics and Science Improvement Consortium (H.M.S.I.C.) and funded through the National Science Foundation Grant #MDR—8319912 to Baylor College of Medicine. Baylor College of Medicine supplied the laboratory space and the materials. We would also like to note our appreciation to all the students who worked with us on this project, especially Phan Duong, who not only did many of the trial experiments but also helped in the preparation of the manuscript.

References

Alcamo, I.E. (1994). *Fundamentals of microbiology*, 4th ed. Menlo Park, CA: The Benjamin Cummings Publishing Co.

Collette, A.T. & Chiappetta, E.L. (1994). *Science instruction in the middle and secondary schools*, 3rd ed. Columbus, OH: Merrill Publishing Co.

Dubos, R. (1962). *The unseen world*. New York: The Rockefeller Institute Press.

Funk, H.J., Okey, J.R., Fiel, R.L., Jaus, H.N. & Sprague, C.S. (1979). *Learning science process skills*. Dubuque, IA: Kendall/Hunt Publishing Co.

Lister, J. (1960). *On the lactic fermentation and its bearing on pathology*. In R.N. Doetsch (Ed.), *Microbiology: Historical contributions from 1776 to 1908*. pp. 76-102. New Brunswick, NJ: Rutgers University Press. (Original work published in 1877).

Nester, E.W., Roberts, C.E., McCarthy, B.J. & Pearsall, N. (1978). *Microbiology: Molecules, microbes, and man*. New York: Holt, Rinehart, and Winston.

Tortora, G.J., Funke, B.R. & Case, C.L. (1992). *Microbiology: An introduction*, 4th ed. Menlo Park, CA: The Benjamin Cummings Publishing Company, Inc.

Pasteurized Milk as an Ecological System

Appendix, Answers to Post-Laboratory Questions

1. Pasteurization is the heating to 163° C to 165° C or 170° C of a substance to kill bacteria. After heating the substance, it is rapidly cooled. No, the harmful bacteria are killed, but other bacteria remain. Bacteria that are heat resistant are left in the milk. Particularly resistant are the species of *Bacillus* that produce endospores.

2. a. cocci b. bacilli c. spirillum; You would expect to find cocci and bacilli in milk.

3. Coagulation is the clotting of liquid milk to form a semisolid. It occurs in milk at room temperature after about three days. It occurs within one day in the incubator and after seven to ten days in the refrigerator.

4. a. temperature, moisture, nutrients and proper atmosphere (oxygen).

4. b. temperature.

5. Milk kept at room temperature spoils in two to three days and at refrigerator temperature in seven to ten days.

6. Quickest to spoil—buttermilk. Slowest to spoil—refrigerated whole milk.

7. Manufacture of yogurt, cheeses, butter and many other dairy products.

8. The pH decreases through time (overall trend).

9. It takes milk two to three days to spoil at room temperature and seven to ten days to spoil at refrigerated temperature.

10. Conclusions. Answers will vary, but some conclusions that may be included are:
    a. Cooling retards bacterial growth in milk.
    b. The pH drops due to bacterial metabolism of lactose to lactic acid.
    c. There is a predictable succession of microorganisms in milk due to changing environmental conditions.

Slime Molds in the Laboratory: Moist Chamber Cultures

Steven L. Stephenson, Fairmont State College, Fairmont, West Virginia

In an article published in *The American Biology Teacher*, February 1982, I described the techniques involved in making field collections of the fruiting bodies of plasmodial slime molds, or Myxomycetes. Slime molds are common inhabitants of decaying plant material throughout the world. However, despite their abundance and widespread occurrence, relatively little use has been made of these fascinating and biologically enigmatic organisms in laboratory studies. There are a number of reasons why this is the case. First of all, because of their small size and the types of situations in which they occur, slime molds are easily overlooked in the field and thus are not familiar organisms to many biology instructors. Moreover, until the appearance of Farr's *How to Know the True Slime Molds* (1981), a fairly comprehensive and relatively nontechnical guide for the identification of slime molds did not exist. Also, the very fact that these organisms do not possess a particularly attractive common name probably hasn't helped matters!

Although slime molds have considerable potential value in laboratory studies, as was pointed out in the previous article, some major constraints do exist in regards to making field collections of these organisms. First of all, since slime mold fruiting bodies tend to be most abundant during summer and early autumn for much of the United States, their period of maximum availability largely falls outside the academic year. Furthermore, many instructors who might otherwise make use of these organisms do not have access to the moist, forested areas where slime mold fruiting bodies are particularly abundant and thus most easily collected. However, laboratory studies of slime molds do not have to be restricted to material collected in the field (or purchased from a biological supply house). The purpose of this article is to describe a technique for obtaining slime molds (plasmodia as well as fruiting bodies) in the laboratory, using pieces of tree bark placed in moist chamber cultures. This technique is relatively easy to carry out and required no special equipment. Moreover, it can be used at any time of the year and in any part of the country.

Background

The use of moist chambers for culturing slime molds was first described by Gilbert and Martin (1933), who placed a few pieces of bark bearing an abundant growth of *Protococcus* in a moist chamber in their laboratory to permit the alga to develop. Much to their surprise, the fruiting bodies of a slime mold also appeared on the bark in the moist chamber. This suggested further examination of similar bark cultures. The two biologists soon discovered that the appearance of slime mold fruiting bodies in such cultures was a common occurrence and that the species obtained in this way included some examples previously thought to be exceedingly rare. Since that time, the moist chamber technique has been used widely to supplement field collections of slime molds. In fact, it is now believed that the bark surface of living trees supports a distinctive group of slime molds (Keller & Brooks 1973), and at least a few species seem to be restricted to this habitat.

Slime Molds

Collecting Bark

The first step in setting up a moist chamber culture involves collecting several pieces ("postage stamp-sized" or smaller) of the dead outer bark of a living tree. Various other types of organic debris, including dead leaves, twigs and the dung of herbivorous animals, also may be used but are generally less productive than bark. Bark from some loose-barked trees may be collected by simply prying off pieces by hand, but for most trees it will be necessary to remove pieces of bark of the desired size carefully using a knife or some other collecting tool. I have found that a small screwdriver is both effective and safe to use. When collecting pieces of bark, care should be taken not to disturb the living tissues of the tree. In addition, pieces of bark that include living tissues, when placed in moist chambers, are often quickly overgrown by filamentous fungi and usually have to be discarded.

Trees with smooth bark are generally regarded as less productive for slime molds than trees with rough bark. Bark from such trees as oak (*Quercus* spp.), ash (*Fraxinus* spp.), elm (*Ulmus* spp.), maple (*Acer* spp.), and hickory (*Carya* spp.) usually produces good results. In general, bark from coniferous trees tends to yield fewer species of slime molds than does bark from broad-leaved trees, although some exceptions do exist. For example, bark from species of juniper (*Juniperus* spp.) is often very productive.

The pieces of bark taken from a given tree should be placed in a single collecting bag and the date, location and type of tree recorded in pencil either on the bag itself or on a slip of paper placed in the bag with the bark. Plastic bags (e.g., sandwich bags) are suitable for short-term storage, but small paper bags should be used if the bark is to be stored for more than a few days.

Preparing Moist Chambers

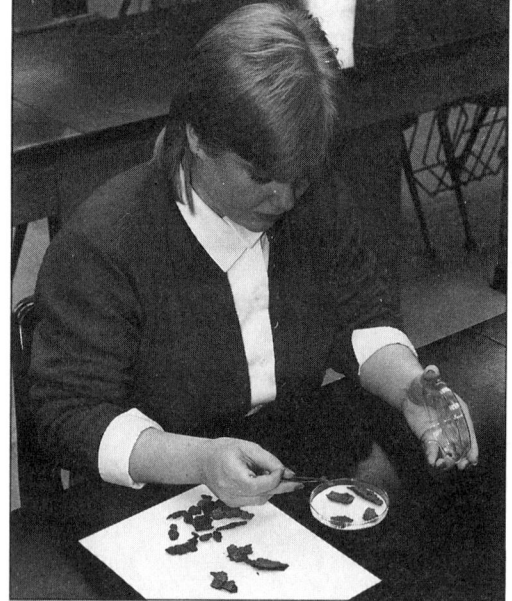

Figure 1. A student prepares a moist chamber culture, using pieces of bark collected in the field.

After collecting the bark, take it back to the laboratory, where the moist chamber cultures are actually prepared. For each culture, fit a petri dish or some other suitable container (e.g., a shallow bowl) with a filter paper disk or a piece of absorbent paper toweling trimmed to the appropriate size. I use disposable plastic petri dishes, which I simply discard when there is no longer any need to maintain the culture. (Plastic petri dishes have one other advantage, which will be mentioned later in this article.) The pieces of bark collected in the field are then placed on the bottom of the paper disk in such a way as to cover as much of the bottom as possible without overlapping and with their cut surfaces facing downward (Figure 1). For laboratory experiments designed to study, for example, the differences that exist in the species of slime molds produced by bark from different types of trees or trees in different habitats, you should take care to prevent the potential contamination of one culture with spores from another by sterilizing all of the equipment used. Handle pieces of bark with forceps which can be flamed or dipped in alcohol before being used to prepare each new culture. The aforementioned procedure is not necessary if the only objective of preparing the cultures is simply to produce a variety of fruiting bodies for study.

After the pieces of bark have been placed in the dish, add enough distilled water to cover the bark completely. If distilled water is not available, ordinary tap water that has been boiled and then allowed to cool may be substituted. Then cover the dish and set aside to allow the bark to soak. Each dish should be labeled, either with a piece of masking tape applied to the side of the dish (so as not to obscure any portion of the bottom of the dish when the latter is viewed from above), or by writing directly on the lid of the dish with a wax pencil. If the latter method is used, it sometimes may be necessary to rotate the lid of the dish when observing the bark in order to view portions of the bottom otherwise obscured by the writing.

The next day, the excess water in the dish should be poured off. After this has been done, set the culture aside and disturb it as little as possible. (If you use petri dishes, you can stack them in order to save space.) Cultures should be kept at room temperatures under normal light conditions. However, they should be placed in a part of the laboratory that does not receive direct sunlight at any time of the day.

Checking the Cultures

Begin observing the cultures a day or two after they have been established, and continue your observations on a fairly regular basis (i.e., at least once every few days) for at least two weeks. After this time, you can check the cultures at less frequent intervals for as long as they are maintained. It is possible to maintain cultures for several months or more. In fact, some species of slime molds that appear in such cultures typically require weeks or even months to develop. In contrast, many of the species that produce very small fruiting bodies (e.g., species of *Echinostelium*, including the very common *E. minutum*) often appear rather quickly, sometimes within a day or two. Since the cultures slowly dry out, it may be necessary to add more water from time to time. Conversely, allowing a culture to dry out completely and then rewetting it often yields species of slime molds that otherwise would never develop.

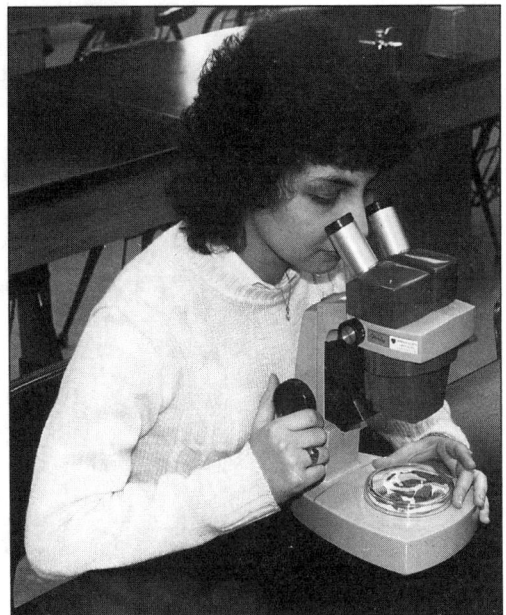

Figure 2. Checking a culture for slime mold fruiting bodies with the aid of a stereo microscope.

It is very important to check cultures as carefully as possible in order to detect all of the fruiting bodies that may be present. Sometimes only one of a few fruiting bodies of a given species will develop; in other cases hundreds of fruiting bodies will be produced. It is not at all unusual for a single culture to yield the fruiting bodies of several different species. Cultures are most

Biology Labs That Work: The Best of How-To-Do-Its

Slime Molds

Figure 3. The fruiting bodies of *Arcyria cinerea*, a slime mold that commonly appears in moist chamber cultures. 5X.

Figure 4. *Perichaena depressa*, a slime mold that is occasionally found in moist chamber cultures. 5X.

easily checked using a stereo microscope (Figure 2). However, if this is not available, you can use a good hand lens (at least 10×). In many instances, the fruiting bodies that develop will be clearly visible to the naked eye. For example, such is the case for the fruiting bodies of *Arcyria cinerea* (Figure 3). Since removing the lid can interfere with the development of fruiting bodies not yet completely mature, the culture should be examined by viewing through the lid whenever possible.

When you detect completely mature fruiting bodies in a culture, harvest them as soon as possible. If this is not done, the fruiting bodies soon become moldy. Fruiting bodies that appear in moist chamber cultures usually develop on the surface of the bark (Figure 4) or on the paper disk at the bottom of the dish. However, fruiting bodies sometimes develop on the side of the dish and may even develop on the lower surface of the lid. When this happens, the only way to harvest the fruiting bodies is to cut away the portion of the dish or lid upon which they developed. Obviously, this is the other advantage of using plastic petri dishes!

Once the fruiting bodies have been harvested, prepare them for storage and further study. In brief, this involves allowing the fruiting bodies and the pieces of bark or paper upon which they developed to air dry, and then placing each individual collection in a small box or tray.

Slime mold plasmodia are often observed in moist chamber cultures and usually provide indirect evidence of their presence in the form of "slime tracks" left behind on the surface of either the bark or paper when not observed directly. Plasmodia which develop in such cultures can be used and studied in the same way as those (e.g., the plasmodium of *Physarum polycephalum*) obtained from a biological supply house. For example, protoplasmic streaming in the main veins of an active plasmodium usually can be observed with the aid of a stereo microscope. Although some plasmodia will persist (i.e., not transform into fruiting bodies) in moist chamber cultures for several days or even longer if favorable conditions are maintained, most eventually do produce fruiting bodies. However, some of the plasmodia that appear in cultures never fruit. When plasmodia fail to produce fruiting bodies, it is sometimes possible to induce fruiting by allowing the culture to dry out completely and then rewetting it several days later. In some instances, increasing the light intensity to which the culture is exposed may help to induce fruiting for those species of slime molds that have yellow or orange plasmodia. Stevens (1981) describes several possible projects involving plasmodia.

Conclusion

Because of its simplicity, the moist chamber technique described in this article can be incorporated successfully into laboratory studies at virtually any academic level. Moreover, for instructors who have never worked with slime molds, it represents a convenient way of becoming familiar with these fascinating organisms while at the same time providing students with a hands-on educational experience that combines field work with an introduction to laboratory procedures.

Slime Molds

References

Farr, M.L. (1981). *How to know the true slime molds.* Dubuque, IA: William C. Brown Company.

Gilbert, H.C. & Martin, G.W. (1933). Myxomycetes found on the bark of living trees. *University of Iowa Studies in Natural History, 15,* 3-8.

Keller, H.W. & Brooks, T.E. (1973). Corticolous Myxomycetes 1: Two new species of Didymium. *Mycologia, 65,* 286-294.

Stephenson, S.L. (1982). Slime molds in the laboratory. *The American Biology Teacher, 44*(2), 119-120, 127.

Stevens, R.B. (Ed.) (1981). *Mycology guidebook.* Seattle: University of Washington Press.

The Almost Ideal Lab--Mutualistic Nitrogen Fixation

Stuart W. Hughes, La Salle University, Philadelphia, Pennsylvania

Of the lab experiences biology teachers conduct, many are observational and skill developing. Others are designed to verify concepts or principles already studied. Still others are guided discovery inquiries with well-established results that are yet unknown to students. Only rarely, in most classrooms, will students be asked to conduct a truly open-ended investigation.

This perhaps is the best educational situation. Students need to develop skills and confidence using equipment and following established techniques before attempting more challenging research. While it seems fashionable to write much about the latter, the practical fact is more classroom-tested labs of the guided discovery type should be made known.

Characteristics of a Good Lab

Most experienced biology teachers would expect a really good lab to provide a controlled experiment with a single variable that would offer students the opportunity to practice determining which tests are controls and which are experimental. The sample sizes would be sufficiently large for the collection of significant data. It would include careful measuring, accurate record keeping, collation and discussion of data with other experimenters, and the chance to analyze the data using statistical techniques.

If the experiment also provides opportunity to use living things over an extended period so students can observe their life history and have individual oversight, it can be a powerful motivator. The superior lab should also lend itself to the retrospective proposal of hypotheses, formulation of deductions and drawing of conclusions as well as the prospect of proposing new questions based on information learned or experienced in the investigation that could lead to new hypotheses and testing.

If the lab experience can be conducted within the constraints of a limited budget, a 45 - 50-minute period, available equipment, and safety concerns and still have a high probability of success in both illustrating the experimental process and a major biological principle, at the student's developmental level, it would seem almost ideal.

There is, of course, no such thing as an ideal lab experience. All labs have limitations. A uniform procedure cannot, by itself, provide much opportunity for individual initiative even though it could spark questions and subsequent individual research. But, of the hundreds of formal lab experiences I have tried, the one herein described comes closest to providing all the above detailed advantages.

Legumes and *Rhizobium*

The mutualistic association between *Rhizobium* bacteria and legumes is a familiar one to

Mutualistic Nitrogen Fixation

all biology teachers. Their important role in the nitrogen cycle and in agriculture form an essential part of the biology curriculum.

No doubt some biology teachers find it useful to have students dig up legumes to examine their nodules. Perhaps some teachers grow a potted bean or clover plant for demonstration of nodules. These are worthy techniques but my belief is that the lab described here is well worth the small additional effort.

I first tried an early version of the following lab as directed in the *Student Laboratory Guide to Biological Science: An Inquiry Into Life*, 1963 edition, 20 years ago. That lab made no provision for careful measurements or collating of data. Moreover, the directions called for sterilization of sand which was used as a growing medium, a rather tedious task. Also, to expect all your students to perform the investigation in school would require much greenhouse space and classroom time. Such difficulties may account for its being dropped from subsequent editions.

An innovation which overcomes the last mentioned difficulty is to have the students grow their plants at home where they can care for them on a daily basis and bring them to class for measurement at the appropriate time of year just prior to their study of the nitrogen cycle. If ecology is a topic that is covered in the months of May and June, the plants grow well on sunny windowsills.

It is important that all materials (except distilled water) be supplied to the students. It is so much easier and certain for the teacher to obtain the required materials in bulk and distribute them than to expect students to locate them. It also ensures uniformity and success in the experiment. The cost for expendables should be no more than about $8 per class.

Experience has shown me that students get great satisfaction when care is taken to explain and demonstrate how to set up the experiment. This is especially true for planting and watering.

Setting Up the Materials

The procedure is to get them started on the experiment six to eight weeks before they will bring their plants to class for measurement. I do not explain the purpose of the experiment. As a guided discovery inquiry, they have weeks of curiosity about the effect of the bacteria on the bean plants. Many students immediately assume that the bacteria will harm the plants. The teacher should assure them that the bacteria are harmless to humans and pets. As a teacher in an urban school system, I have never had a student who was familiar previously with the inoculation of beans as practiced routinely by farmers (or at least expressed it). The only information I give the students is that the bacteria normally invade the roots of the plant, and swelling or nodules then form. Our purpose is to assess the effect of the bacteria on the plants. They are also told that the way we will determine the effect is to measure the plants' weight and height.

Mutualistic Nitrogen Fixation

Part of two class periods should be scheduled to get the students started. About 15 minutes is needed during the first period to introduce the students to the investigation, answer questions and prepare them for the next period when they receive two 5-inch plastic pots and enough horticultural vermiculite (preferred—less dusty) or perlite for two pots. They are asked to bring a deposit (perhaps $1) for the pots. This deposit will be returned to them when the experiment is completed and they return the cleaned pots. If this deposit is not obtained, the annual loss of pots may be prohibitive. The only significant cost of this activity is the initial outlay for the pots—a one-time expense.

It is essential that the pots be "sterile" in the sense that they be free of *Rhizobium* and fungus spores or any free-living, nitrogen-fixing microbes when distributed to the students. This is no problem if the pots are new; thereafter they can be washed in a household bleach solution before thorough rinsing and drying. Vermiculite has the advantage of being inert and sterile. It may be purchased in large quantities from garden supply stores.

Students may be given any type of legume seeds, but bush type lima beans of high viability (previous year's crop) have proven quite satisfactory. It was suggested for the original BSCS lab that the seeds be prepared by treating them with a fungicide. Teachers might want to try this but my experience has been that if all other materials are "sterile" only a few students have problems with fungus. I find it convenient to have them wrap the seeds up in half a piece of notebook paper. The other half piece may be folded to receive the granulized commercial *Rhizobium* preparation. This is purchased from most seed companies or garden supply stores, or obtained free for educational purposes. It is vital to emphasize that care be taken to prevent the *Rhizobium* from contaminating the seeds, vermiculite or pots.

Tape may be put over the pot holes and punctured for drainage after the first watering lest the vermiculite pour out the holes. The students are given the following instructions to be followed at home:

- Purchase a gallon of distilled or deionized water, usually obtainable from supermarkets. (Caution students about how it differs from bottled spring water. Tap water may be used, but some loss of control over the composition of nutrients provided to the plants would result. This might not affect the final data but it is important in instructing how to conduct a well controlled experiment. No student should feel compelled to obtain the distilled water. Most will try because they understand the importance of taking such care.)

- Label each pot, perhaps with masking tape, one with the word "with" and the other "without."

- Fill each pot with an equal amount of the vermiculite.

- Plant three seeds into each pot equidistant from each other to a depth appropriate for the seed type (about 2 cm below the surface for limas).

Mutualistic Nitrogen Fixation

- Inoculate the "with" seeds with *Rhizobium* just prior to planting. (One could also withhold about one half of the vermiculite to be poured over the seeds after they are positioned and inoculated. The students must be cautioned to work with the "with" pot only after the "without" is prepared lest they accidently inoculate the "without" seeds.)

Note: I have found that students are much more likely to set up the experiment properly if I take the time, on the day prior to distribution of materials, to demonstrate to each class labeling of a pair of pots, adding vermiculite and seeds, and then inoculating the "with" pot with *Rhizobium* bacteria. I then grow the plants under Gro-Lux® lamps for six to eight weeks in class. On the day before the students are to bring their plants to class, I again demonstrate the proper way to prepare them using the plants I have grown. My plants are kept in the refrigerator overnight and can be measured by the occasional student who forgets to bring his or her own. About a week after distribution of seeds students may be given N-free nutrients, but the initial watering may be done with distilled (or tap) water. The most difficult instruction to give involves proper watering. Most 10th grade students have never grown their own potted plants.

- To properly water potted plants, students should water until water comes out the drainage holes.

- Cover the pots with clear plastic wrap and do not water again until the seeds have germinated.

- Keep the pots in a safe, warm place.

Growing the Plants

In northern areas, spring nights may be quite cold so it would be unwise to put the pots on the windowsills until germination has occurred. Even then, removing them from the windowsills after dark and replacing them in the morning will accelerate their growth and discourage fungal growth. Since most of the early growth will result from food stored in the cotyledons, providing N-free nutrients will be unnecessary until they have achieved their first true leaves. At that time the students should add about 1.3 grams of N-free nutrients to each liter of the remaining distilled water.

About a level teaspoonful should be distributed to each student. The proportions of each ingredient are indicated in Table 1. Make 200-400 times these amounts to accommodate several classes of students. If kept in a wide-mouthed, tightly capped container it will provide easy access, and the powder (which must be thoroughly mixed) will not absorb water vapor and harden. Careful measurement of the amount distributed to the students is not essential; they should ensure that equal amounts of the solution are added to each pot when watering appears to be needed.

Mutualistic Nitrogen Fixation

Table 1. Nitrogen-free nutrient proportions*.

0.8 g potassium monohydrogen phosphate (K_2HPO_4)
0.2 g potassium dihydrogen phosphate (KH_2PO_4)
0.2 g magnesium sulfate ($MgSO_4 \cdot 7H_2O$)
0.1 g calcium sulfate ($CaSO_4$)
0.01 g ferric sulfate ($Fe_2(SO_4)_3$)

*From *Teacher's Manual for Student Laboratory Guide, Biological Science: An Inquiry into Science*, BSCS, 1963, Harcourt, Brace and World, New York, p. 277.

Note: While this formulation provides iron needed by *Rhizobium* to incorporate into the enzyme nitrogenase, essential for N-fixation, it lacks molybdenum needed for the molybdoferredoxin part of the enzyme. The easiest way to provide the proper amount is to use *Rhizobium* inoculant which contains Mo in trace amounts.

Commonly, not every seed in each pot will germinate. Inform the students not to worry unless none emerges from either pot by the end of one week. The occasional student with such a problem may start over with new seeds and vermiculite but should sterilize the pots to remove any fungus.

Students should also be cautioned to explain the experiment fully to their parents and ask for the use of the sunniest available windowsill with equal light for each pot. Precautions should also be taken to avoid the hazards of family pets and younger siblings.

Frequently the plants grow so tall they start to fall over and must be supported carefully with thin sticks or by taping the stems to the window, etc. Most students achieve blooms on their plants and many produce at least small pods. It is encouraging for them to observe the growth, flowering and fruiting of their own plants.

Preparing for Class

The day before the plants are to be measured in the lab, the students must prepare the plants for transport to class as follows:

1. Mark a piece of paper or tape for each plant with appropriate labels, "with" or "without" and attach to the stems.

2. Obtain a plastic bag, plastic wrap or aluminum foil for wrapping the plants.

3. Bring the plants, plastic bag and a bucket of water outside the house into a garden, or near foundation plants. A water hose will also work.

4. Carefully remove the whole rootmass from the pot and gently shake off the vermicu-

Mutualistic Nitrogen Fixation

lite. Take care to save all the roots and nodules. To get all the vermiculite off, dip the roots up and down in the water. Squeeze out the water and put the plants in the bag. Do not attempt to separate the plants but they may be rolled or bent. Note that the vermiculite is good for the garden soil. This operation may be done indoors, however care must be taken to prevent the vermiculite from going down the drain. Vermiculite may be sieved out of a bucket of water.

5. Repeat with the other plants. Take care that the plants are wrapped and do not dry out. Place in a school bag or in a place where you will not forget to bring it to the next class.

Collecting Data

Classroom measurements are straightforward. The heights (lengths) of the stems are measured from the points of attachment to the roots to the growing tips (see Figure 1). Round off to the nearest centimeter. The lengths of all plants in a pot are summed and recorded. Similarly, all plants in each pot are weighed to the nearest gram (Figure 2) and recorded. As students finish their measurements they may record them on the chalkboard or report to the teacher for recording. The number of plants in each pot should also be recorded so that factor may be properly adjusted.

At the end of the measurement period, all plants are collected in a box. Another important lesson may be taught if you emphasize that such valuable organic matter is best used in the garden as a mulch or composted. If it were put in the trash, it might be incinerated and pollute the air.

Figure 1. Measuring the stem length of bean plants.

An example of some data is shown in Table 2. The figures have been adjusted to the actual number of plants counted. For example, if 100 versus 96 were counted, the first total was multiplied by 0.96. Note that in Class #3, the length of the "without" plants actually exceeded the length of the "with" plants, although the mass of the "with" plants was greater by seven grams. In cases like this it is interesting to look for nodules on the "without" plants. Sometimes the degree of difference between experimental and control groups more closely

Mutualistic Nitrogen Fixation

Figure 2. Weighing the plants can be fun.

matches the ability grouping of the students than any single factor. A lack of significant difference in one class can be turned into an opportunity to again illustrate the importance of a large sample size.

Data Analysis

Prior to tabulation of class data it is important to ensure that a student's data not be included if nodules have formed on the control plants. This would indicate contamination during inoculation. While the student is not penalized, all students must understand why those data cannot be used. Once the data from all the students' plants have been measured and summed, there should be as careful an analysis as time and student abilities will allow. The students may be asked how they might evaluate the measured

Table 2. Total data from four classes of students.

	Plant Weight (nearest g)		Plant Length (nearest cm)	
	With *Rhizobium*	W/O *Rhizobium*	With *Rhizobium*	W/O *Rhizobium*
Class 1	437	410	1113	1037
Class 2	637	539	2170	2012
Class 3	293	286	1779	1793
Class 4	271	269	1229	1117
Totals	1638	1504	6291	5959

Mutualistic Nitrogen Fixation

differences. It will usually be suggested that we should calculate the percentage increase in weight and height. Thus the differences in the totals for the four classes of students of about 100 sets of plants were:

	Weight	Length
"With"	1638 grams	6291 centimeters
"Without"	1504 grams	5959 centimeters
Difference	134 grams	232 centimeters

The percentage increase in weight would be 134/1504 = 0.089 or 8.9%. The percentage increase in height would be 232/5959 = 0.039 or 3.9%.

These figures should provoke further inquiry, preferably student generated. Examples might be:

- Is weight more important than height (or vice versa)?

- What is the ultimate goal in growing beans? Is it to maximize plant weight and height or fruit or seeds?

- If we cannot easily measure fruit or seeds, should we value the weight measurement more than the height measurement?

- Eventually, students may ask whether or not a 4-9% increase might not be due to chance (i.e., Is the increase significant?).

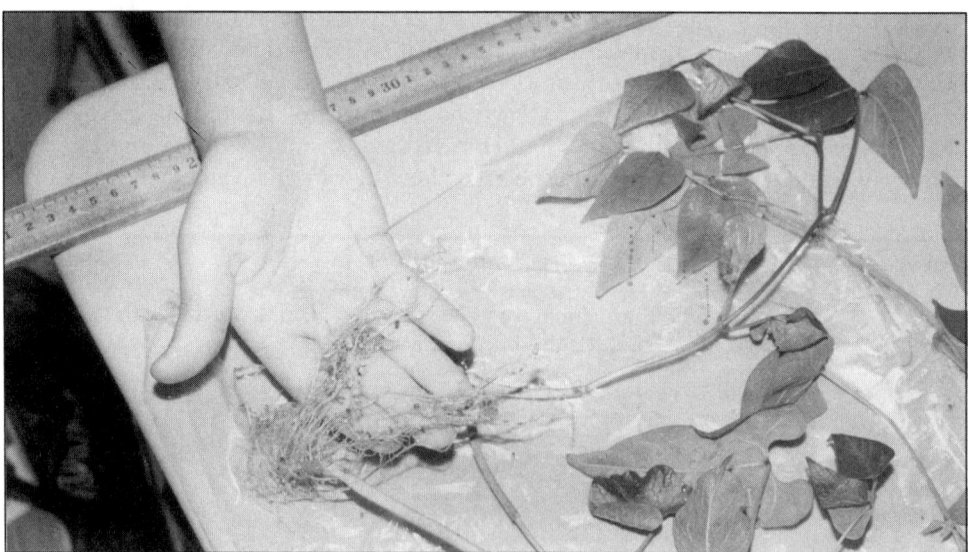

Figure 3. Nodules on the roots of "with" plants.

Many teachers may feel that if students are guided to this question this experiment will have fulfilled our expectations. Some statement about a 5% increase as *usually* being significant would be instructive and would not necessitate further class time. However, for the above average class or if the teacher feels a statistical analysis is appropriate, several tests are available. At this late point in the school year (spring), such analysis may have already been used with other experiments and the teacher may wish to take no further class time and may allow only a few students to do it on their own time. If a classroom analysis is to be made, choice of the statistical test will depend on such variables as the teacher's interest, familiarity with the test, etc. A t-test could be used for these data since what is being asked is whether the two samples ("with" and "without" plants) are from the same or different populations. Figure 4 gives the formula for the t-test. Any statistics book would provide further details. The BSCS *Biology Teacher's Handbook* (any edition) also explains its use. Unfortunately, it requires tedious arrays of the individual plant measurements to generate standard deviations. I feel this is impractical and too time-consuming in the biology class. Students could lose perspective of the primary goal of learning the experimental process.

Mutualistic Nitroger Fixation

$$t = \frac{\bar{x}_1 - \bar{x}_2}{s\sqrt{1/n_1 + 1/n_2}}$$

where x_1 is the mean of sample 1, x_2 is the mean of sample 2, and n_1 and n_2 the numbers of individuals in each sample. The sample standard deviation, found by combining the data for two samples, is designated s and calculated:

$$s = \sqrt{f_1 s_1^2 + f_2 s_2^2 / f_1 + f_2}$$

where $f_1 = n_1 - 1$ and $f_2 = n_2 - 1$, and s_1 and s_2 are standard deviations of *each sample*. In the t-test the number of degrees of freedom is $f_1 + f_2$.

Figure 4. t-test of significant difference.

An alternative approach would be to use the χ^2 (chi square) test. Some may question its precise applicability to a 2-value test, but it has several advantages. BSCS teachers are familiar with the test. By the end of the school year the students may have already used it in analysis of *Drosophila* dihybrid cross data and it would then be reinforced. It is less time-consuming and the average student

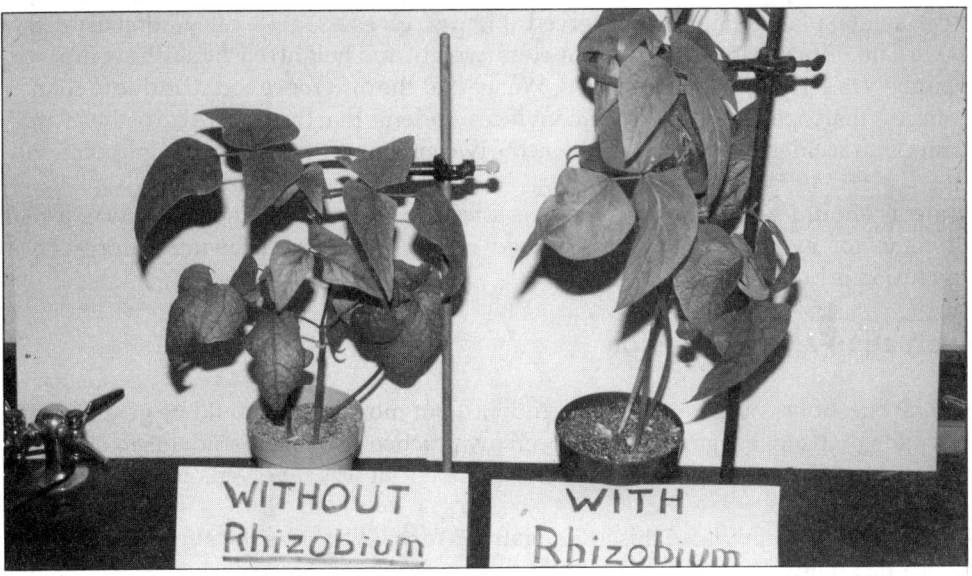

Figure 5. Not all plants show this clear difference in height.

Biology Labs That Work: The Best of How-To-Do-Its

Mutualistic Nitrogen Fixation

$$\chi^2 = \Sigma \frac{(e-o)^2}{e}$$

For weight: $\chi^2_w = \frac{(1638-1504)^2}{1638}$ For length: $\chi^2_l = \frac{(6291-5959)^2}{6291}$

$\chi^2_w = 10.9$ $\chi^2_l = 17.5$

where Σ = sum of all
e = expected ("with")
o = observed ("without")
w = weight
l = length (height)

Chi-square for two classes	0.0002	0.004	0.455	1.074	1.642	2.706	3.841	6.635
The number of times in 100 that chance alone could be responsible for the variation	99	95	50	30	20	10	5	1

Figure 5. $\chi 2$ (chi square) using data from Table 2.

finds it much less formidable. Again, we want students to succeed and not be repelled by science. The test works by simply using the data for the control (situation in nature, i.e., the "with" plants) as the expected, and the modified or unnatural, "without" experimental plant data, as the observed. Figure 5 gives a sample calculation using the above data. Since there are two parameters (weight and height) each may be compared separately to a χ^2 table for two classes. We see that the difference could be due to chance alone less than one time in 100. This convinces students that the larger size of the "with" plants is no accident; in statistical jargon, the two groups are from different populations.

Students should be cautioned that such a test does not *prove* that there was a real difference or even that *Rhizobium* caused it—only that the probability for such a conclusion is high.

Evaluation and Discussion

Several questions could be asked the students, but most credit should be given for the execution of the investigation. It has been my practice also to ask students to describe, in their reports, each step of the scientific method that was involved. For example:

- What question stimulated this investigation? (ans. What was the effect of *Rhizobium* bacteria on lima bean plants?)

Mutualistic Nitrogen Fixation

- What hypothesis was proposed to answer that question? (ans. *Rhizobium* helps the lima bean plants grow larger; or hinders or has no effect.)

- What deduction from the hypothesis led to the experiment? (ans. If *Rhizobium* helped the plants, then by withholding the *Rhizobium* the plants should not grow as well as plants with it).

- What conclusion is drawn concerning the experimental test of the hypothesis? (ans. *Rhizobium* appears to help lima bean plants grow larger.)

- What new questions could you ask that seem testable? This last question, of course, could lead to some open-ended inquiry which the motivated student may be encouraged to pursue. Some examples are:

  1. What do *Rhizobium* bacteria look like under the microscope?

  2. Can the *Rhizobium* be isolated and grown in pure culture?

  3. Are certain species of *Rhizobium* specific for certain legume species? Several species are readily available from supply houses.

  4. How does the *Rhizobium* enter the roots?

  5. How much nitrogen (nitrate fertilizer) would be needed to equal the effect of *Rhizobium* ?

The first question is most easily investigated and was, in fact, included in the original BSCS lab. The others are increasingly difficult but are possible projects for interested students.

Further class discussion of the vigorous ongoing research in many laboratories around the world into ways of understanding and improving mutualistic nitrogen-fixation, including recombinant DNA studies, seems worthwhile. The completion and discussion of this experiment makes study of the nitrogen cycle more meaningful to students, but, more importantly, enables them to enjoy a guided discovery experience in experimental biology. I believe if many biology teachers gave it an honest try, they would also find it to be an almost ideal lab.

Quantifying Intracellular Water Regulation in a Single-Celled Organism

Barton L. Bergquist, University of Northern Iowa, Cedar Falls, Iowa

Homeostatic regulation of biological fluids, one of the most important concepts in understanding living system functions, has been demonstrated and taught using varied biological models; e.g. dandelions (Bergquist 1981), erythrocytes (Parsons & Schapiro 1975), oyster-plant and potato tubers (Machlis & Torrey 1956). Yet, it is sometimes difficult to teach this idea as a dynamic concept, rather than as factual statements to be memorized. This article describes a relatively simple laboratory experiment that demonstrates the influx and subsequent removal of cellular water by the common ciliate *Tetrahymena*. Students will gain experience in quantitative biological experimentation and observational and descriptive skills.

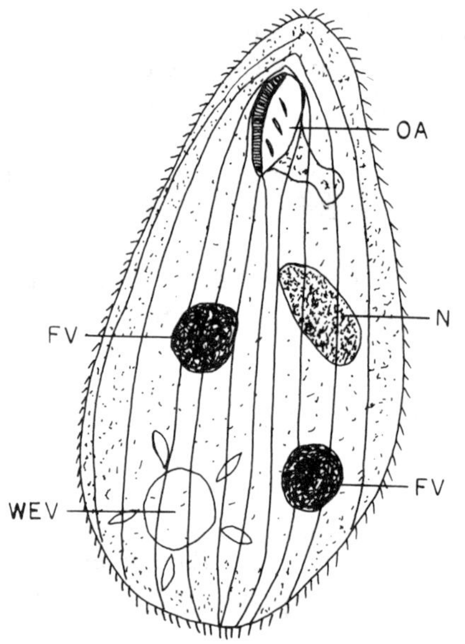

Figure 1. Structure of *Tetrahymena*. FV-food vacuole, N-nucleus (Macronucleus), OA-oral apparatus, CV-contractile vacuole.

The techniques described are designed to allow a class of students to collect data, then analyze the data by graphical and statistical descriptions regarding the relations within the experiment. The example depicted includes several descriptive-analytical techniques for use as desired. This exercise is directed toward college level undergraduates.

Tetrahymena, a single-celled protozoan, is a relative of the slightly larger *Paramecium*. While widely found in nature (Corliss 1973), it is also commonly used in biological experimentation (Hill 1972; Elliot 1973; Bovee 1979). The general structure of *Tetrahymena* is illustrated in Figure 1. The spherical contractile vacuole (CV) is located near the cell membrane in the posterior portion of the organism. This structure is also known as the contractile vacuole, but CV more adequately describes its function. The CV slowly expands as it accumulates water from the cytoplasm of the cell and then suddenly contracts to expel its contents outside the cell. The process is regulated precisely, giving good data for study and analysis.

The water regulating function of the contractile vacuole in protozoans has been reviewed (Conner 1967; Kitching 1967; Patterson 1980; Van Rossam et al 1987) and its role in ionic and osmotic regulation in *Tetrahymena* is well documented (Dunham & Kropp 1973).

Tetrahymena is hyperosmotic to fresh water, so that continual entry of water occurs in response to the external solute concentration. In fact, water entry may not be based only upon the classical idea of osmosis, since experimental evidence has suggested that both the "Gibbs-Donnan ratio" and sponge-like imbibition of water by cytoplasmic proteins are involved (Organ, Bovee & Jahn 1978). However, in this study, *Tetrahymena* will be placed in phosphate buffer solutions of differing molarities and, as the inward water flux varies, so too will the need to control the intracellular fluid volume. *Tetrahymena* expels excess water by contraction of its CV (Dunham & Kropp 1973). Functionally, we can liken the CV to the multicellular organism's urinary bladder because it stores liquid waste products until it contracts and discharges its liquid contents.

Quantifying Intracellular Water Regulation

Using one of the methods described here, one can retard the motility of the cells and then count the rate of CV contractions. The buffer concentration, or osmolarity, can then be compared with the rate of contraction to illustrate the relationship between the two.

Materials and Methods

Supplies for each student or student group

- Five buffer solutions (15 ml each). Solutions of 0.066 M KH_2PO_4 and 0.066 M K_2HPO_4 are added together, stirring until pH 6.8 is reached (monitor with a pH meter). The stock (pH 6.8) solution is diluted with distilled water to the desired concentrations over the range of 0.002 M to 0.0166 M.

- Petroleum jelly.

- Two or three glass slides and cover slips per buffer solution tested.

- Compound microscope with 200✕-400✕ magnification.

- Conical centrifuge tubes (12-15 ml), one for each buffer solution plus a balancing tube.

- Five pipets (or one that is washed between buffers).

Supplies for Common Use

- *Tetrahymena* in pure culture (may be obtained through biological supply houses). Note: *Tetrahymena pigmentosa* from the author's own laboratory stock was used in the illustrated experiments. Other species of *Tetrahymena* are also acceptable.

- Clinical centrifuge.

- Measuring pipet (5-10 ml).

Quantifying Intracellular Water Regulation

Procedure

1. Place a small glob of petroleum jelly in the center of a clean slide. The key here is to get sufficient jelly in which to entrap organisms, but not so much as to create a mess too thick to be usable. Similar methods of entrapment have been described by Organ et al (1968) and Organ et al (1978).

2. Spread the jelly to form a small raised spot, about 1-2 cm in diameter, on the slide. Make the surface texture of the jelly spot uneven by alternately touching and withdrawing your finger upward away from the slide.

3. Pipet a small volume of cells (1-5 ml, depending on cell density) from the cell culture into a conical centrifuge tube.

4. Gently centrifuge the cells into a pellet (approximately 1.5 minutes at intermediate speed).

5. Carefully decant the supernatant to leave the cells relatively "dry."

6. Resuspend the cells in 5 ml or more of the buffer to be tested by vortexing or dislodging with squirts from a pipet equipped with a rubber bulb. If the cells have been removed adequately from their culture medium and resuspended in sufficient testing medium, one centrifugation and resuspension is adequate. For more carefully controlled experiments, a second centrifugation and resuspension is recommended.

7. Pipet a drop of the cell culture from the bottom of the centrifuge tube into the middle of the petroleum jelly. Be careful not to get too much fluid.

Table 1. Raw data--illustration from 0.0166 M phosphate solution.

Cell #	Contractions per 5 min.	Rate per min.	Cell #	Contractions per 5 min.	Rate per min.	Cell #	Contractions per 5 min.	Rate per min.
1	13	2.6	11	10	2.0	21	14	2.8
2	11	2.2	12	14	2.8	22	11	2.2
3	10	2.0	13	13	2.6	23	11	2.2
4	11	2.2	14	15	3.0	24	10	2.0
5	14	2.8	15	10	2.0	25	15	3.0
6	12	2.4	16	11	2.2	26	15	3.0
7	15	3.0	17	14	2.8	27	10	2.0
8	13	2.6	18	12	2.4	28	11	2.2
9	11	2.2	19	13	2.6	29	12	2.4
10	12	2.4	20	11	2.2	30	15	3.0

30 cell Composite Average   2.46
30 cell Composite Std. Dev.   +/-0.353

8. Place a cover glass over the spot and gently press it down to squash the jelly and trap the cells in pockets within the jelly. Be careful not to overdo this process; this step may require practice.

Note: An alternate trapping technique can be accomplished by using a small piece of cotton. A dense matrix of the fibers will create enough small cavities in which *Tetrahymena* can be isolated. Care must be taken to use a relatively small amount of cotton which has been flattened on the slide. Chemicals that slow the cells must be avoided to prevent artifacts.

9. Place the slide on the microscope stage and search for cells that are entrapped in a small reservoir of buffer, but are still intact and alive.

10. Find a typical medium size cell and observe the behavior of the CV (Figure 1).

11. Count the number of contractions of the CV over a period of five minutes and record your results in contractions/minute.

Table 2. Composite data.

Experimental Solution	Sample Size (N)	Total Counts	Rate/Min.	Std. Dev.	t-Test Values for pairs	Probability
1. 0.00208 M	30	882	5.88	0.402		
					-1.95	0.056
2. 0.04166 M	30	851	5.67	0.418		
					-8.95	0.0001
3. 0.0833 M	30	678	4.52	0.593		
					-7.15	0.0001
4. 0.0125 M	30	517	3.44	0.532		
					-8.46	0.0001
5. 0.0166 M	30	369	2.46	0.353		

Table 3. Analysis of variance table.

Source	Sum Sqs.	DF	Mean Sqs.	F	P
Between	254.735	4	63.68	290.31	0.00001
Within	31.808	145	0.22		

Linear Regression Equation: $Y = A + (B \times X)$
= Contraction Rate = Intercept + (Slope × Molarity)
= Contraction Rate = 6.5247 + (-244.45 × Molarity)

Quantifying Intracellular Water Regulation

12. Determine the rate for several cells in each buffer solution. A single slide preparation can be used for several replicates with each experimental solution. Avoid heating by the microscope's light; use minimal light intensity.

13. Repeat using other experimental solutions (0.002 to 0.0166 M or other similar concentrations) and record your data. The class may be subdivided into groups, each using different molar solutions.

14. Use the t-test to identify rates that are significantly different from each other. Other statistical tests, as illustrated in Tables 2 and 3, may also be used in describing the data.

Treatment of the Results and Discussion

The illustrated experiment may be analyzed using one or more of the following techniques: graphics (Figure 2); paired comparison of groups using the t-test (Table 2); or analysis of variance and regression (Table 3). A detailed description of the statistics may be found in standard references on the subject (Snedecor & Cochran 1980; Sokol & Rohlf 1981; Zar 1984). Use of these analysis techniques will vary according to the level of the student investigators and purpose involved.

Since the CV contraction rate depends on buffer concentration, it is appropriate to describe this relationship using a regression; the correlation coefficient (r) would not be suitable (Sokol & Rohlf 1981).

The data show a linear relationship between the concentration of the phosphate buffer solution and the CV contraction rate (Figure 2). However, nonlinearity occurs at lower molarity. Other experimental data have also shown that *Tetrahymena* changes CV output in a nonlinear manner at more extreme molarities (Stoner & Dunham 1970). Therefore use of extreme concentrations in this experimental protocol may also lead to a nonlinear relationship. Other solutes (e.g., sucrose, calcium, potassium, sodium, heavy metals or combinations of these) might also be usable in alternate types of investigations.

These data were obtained at a temperature of approximately 22°C. Temperature might be a variable to consider for other experiments, but must be held constant in the experiment as defined above.

A class of 20-30 students may be subdivided, assigning different ionic concentrations to different groups of students. Like data are then pooled into common groups for analysis.

Variations of this experiment may also be developed as individual student science fair or class projects.

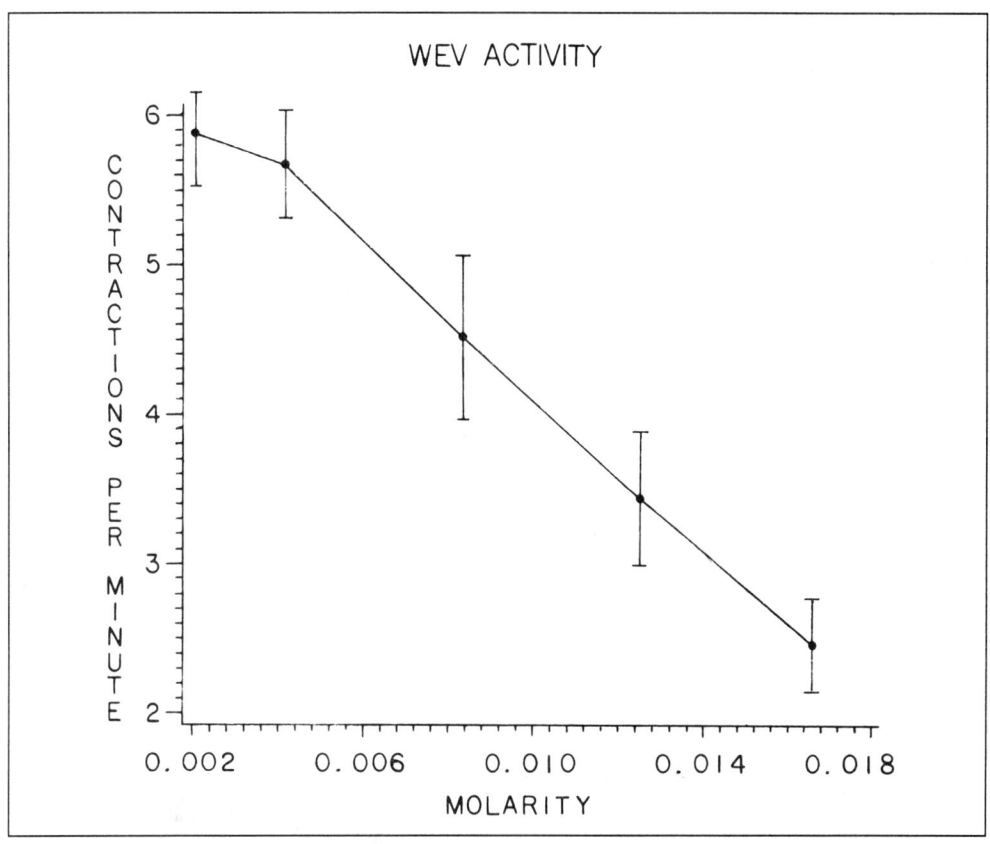

Figure 2. CV contractile activity in relationship to the molarity of the buffer solutions. Standard deviations for the mean at each tested molarity are indicated by vertical bars.

Acknowledgment

The technical assistance of Brian Boyd and manuscript review by Dr. Alan Orr are gratefully acknowledged.

References

Bergquist, B.L. (1981). Dandelion floral stems: A model for teaching intercellular tonicity. *The American Biology Teacher*, 43(1), 45-47.

Bovee, E.C. (1979). *Tetrahymena*: Cell for all seasons. *Iowa Science Teachers Journal*, 16(2), 12-17.

Conner, R.L. (1967). Transport phenomena in protozoa. In M. Florkin & B.J. Scheer (Eds.), *Chemical Zoology, Vol. 1, Protozoa* (pp. 309-350). (G.W. Kidder, Ed.). New York: Academic Press.

Quantifying Intracellular Water Regulation

Corliss, J.O. (1973). History, taxonomy, ecology, and evolution of species of *Tetrahymena*. In A.M. Elliot (Ed.), *Biology of* Tetrahymena (pp. 1-55). Stroudsburg, PA: Dowden, Hutchinson and Ross, Inc.

Dunham, P.B. & Kropp, D.L. (1973). Regulation of solutes and water in *Tetrahymena*. In A.M. Elliot (Ed.), *Biology of* Tetrahymena (pp. 165-198). Stroudsburg, PA: Dowden, Hutchinson and Ross, Inc.

Elliot, A.M. (1973). *Biology of* Tetrahymena. Stroudsburg, PA: Dowden, Hutchinson and Ross, Inc.

Hill, D.L. (1972). *The biochemistry and physiology of* Tetrahymena. NY: Academic Press.

Kitching, J.A. (1967). Contractile vacuoles, ionic regulation and excretion. In T.T. Chen (Ed.), *Research in Protozoology*, Vol. 1. (pp. 307-336). NY: Pergamon Press.

Machlis, L. & Torrey, J.G. (1956). *Plants in action: A laboratory manual of plant physiology*. San Francisco, CA: W.H. Freeman and Company.

Nematbakhsh, S. & Bergquist, B. (1993). Periodicity and the influence of temperature and cellular size in contractile vacuole contraction intervals. *Transactions of the American Microscopical Society, 112*(4): 292-305.

Organ, A.E., Bovee, E.C. & Jahn, T.L. (1978). Effects of ionic ratios vs. osmotic pressure on the rate of the water expelling vesicle of *Tetrahymena pyriformis*. *Acta Protozoology, 17*(1), 177-190.

Organ, A.E., Bovee, E.C., Jahn, T.L., Wigg, D., & Fonesca, J.R. (1968). The mechanism of the nephridial apparatus of *Paramecium multi-micronucleotum*. *Journal of Cell Biology, 37*, 139-145.

Parsons, J.A. & Schapiro, H.C. (1975). *Exercises in cell biology*. NY: McGraw-Hill Book Company.

Patterson, D.J. (1980). Contractile vacuoles and associated structures: Their organization and function. *Biological Reviews and Biological Proceedings of the Cambridge Philosophical Society, 55*: 1-46.

Snedecor, G.W. & Cochran, W.G. (1980). *Statistical methods*. Ames, IA: The Iowa State University Press.

Sokol, R.R. & Rohlf, F.J. (1981). *Biometry*. San Francisco, CA: W.H. Freeman and Company.

Stoner, L.L. & Dunham, P.B. (1970). Regulation of cellular osmolarity and volume in *Tetrahymena*. *Journal of Experimental Biology, 53*, 391-399.

Van Rossam, G.P.V., Russo, M.A. & Schisselbauer, J.C. (1987). Role of cytoplasmic vesicles in volume maintenance. *Current Topics in Membranes and Transport, 30*: 45-74.

Zar, J.H. (1984). *Biostatistical analysis*. Englewood Cliffs, NJ: Prentice-Hall.

Using Yeast and Ultraviolet Radiation To Introduce the Scientific Method

Lois T. Mayo, Pius X High School, Lincoln, Nebraska
Pat J. Friedrichsen, Lincoln High School, Lincoln, Nebraska

During the first week of school, time in secondary school science class is frequently spent discussing the tools and techniques used by scientists. Yager (1988) states that one of the pitfalls of current teaching methods is that science teachers spend much time preparing students to learn the rules and skills of science without actually allowing them to do science. "Like athletes, science students may need to play the game frequently, to use the information and skills they possess, and to encounter a real need for more background and more skills" (Yager 1988).

We designed a laboratory activity for use during the first or second week of school that engages the students in the basic principles of the scientific method. We wanted a scientific question that was both meaningful and relevant to our students. Such an issue exists with the dramatic increase in the rate of skin cancers among young people. The American Cancer Society (1992) estimates that there are more than 600,000 new cases of skin cancer every year, making it the most common form of human cancer. From news stories and previous studies, most students know that ultraviolet (UV) radiation is found in sunlight and damages cells. They have heard about the connection between UV radiation and skin cancer. Thus, most students have at least some prior knowledge and interest in this subject.

Before performing an experiment, we asked our students to consider what they know about UV radiation and its influence on living cells and to write a hypothesis to predict the outcome.

The students then conduct a simple experiment in which half of a yeast population is exposed to an experimental factor (UV radiation) for exactly 20 seconds, and half is not exposed. The beauty of this experiment is that the control and experimental treatment are side by side in the same petri dish. After analyzing class data, students are asked to accept, reject or modify their hypothesis. Having performed an actual experiment, students are able to participate in a more meaningful discussion of this scenario of the scientific method.

For this experiment, we used a strain of yeast, *Saccharomyces cerevisiae*, which is sensitive to UV radiation (Manney & Manney 1990). It is an ideal organism to use in student labs. The yeast is easy to grow, safe to handle and requires little space. Its growth rate can be manipulated by varying the temperature. The yeast culture is easy to maintain from one year to the next, eliminating replacement costs.

Materials

- *Saccharomyces cerevisiae* culture, adenine 2, strain HA2

- Petri plates containing yeast extract

Yeast and UV Radiation

- Dextrose agar
- UV radiation source
- Incubator (optional)
- Glass spreaders
- 95% ethanol
- Flame source
- Index cards

Note: The index cards may be autoclaved and wrapped in foil to reduce contamination.

Procedure

1. Sterile techniques must be used for the entire experiment since the space outside the UV box is not sterile, and contamination must be avoided. The inside of the UV box should be relatively sterile because it is enclosed and it is exposed to UV radiation. Test tubes should be flamed whenever their covers are removed. Petri plate lids should be removed for the shortest time possible.

2. Collect a small amount of yeast onto a sterile loop and place in 1 ml sterile, distilled water using just enough yeast to make the water slightly turbid.

3. Make a standard serial dilution of the yeast using four test tubes in a series as described by Manney and Manney (1990).

4. Inoculate an agar plate with 0.1 ml of diluted yeast solution from the fourth test tube in the series. Spread the yeast solution with a glass spreader sterilized by dipping in 95% ethanol and flaming.

5. Remove the petri lid and cover half of the plate with an index card (Figure 1). It is important to do this, since UV radiation cannot penetrate plastic or glass. (The portion of the plate shielded from the UV radiation with the index card will be the control. The part exposed to the UV radiation will show the results of the experimental treatment.)

6. Expose the plate to UV radiation for 20 seconds using a UV radiation source. Variable times could be used for additional studies and comparisons. Remove the plate and replace the lid.

Table 1. Total number of yeast colonies.

Class #	No UV Exposure	20 Sec. UV Exposure
1	213	70
2	943	260
3	277	175
4	369	106
5	347	94
6	424	256
Total	2573	961
Average[1]	429	160

[1]The obvious difference in the means of the treated and untreated samples was statistically significant by the t-test ($p < 0.05$).

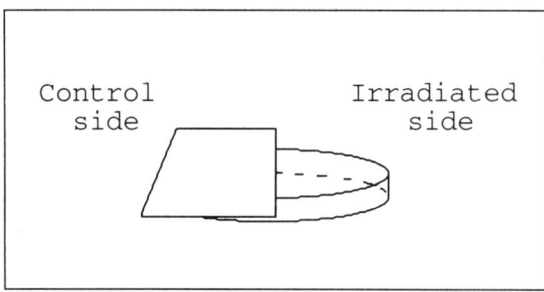

Figure 1. Experimental setup.

Yeast and UV Radiation

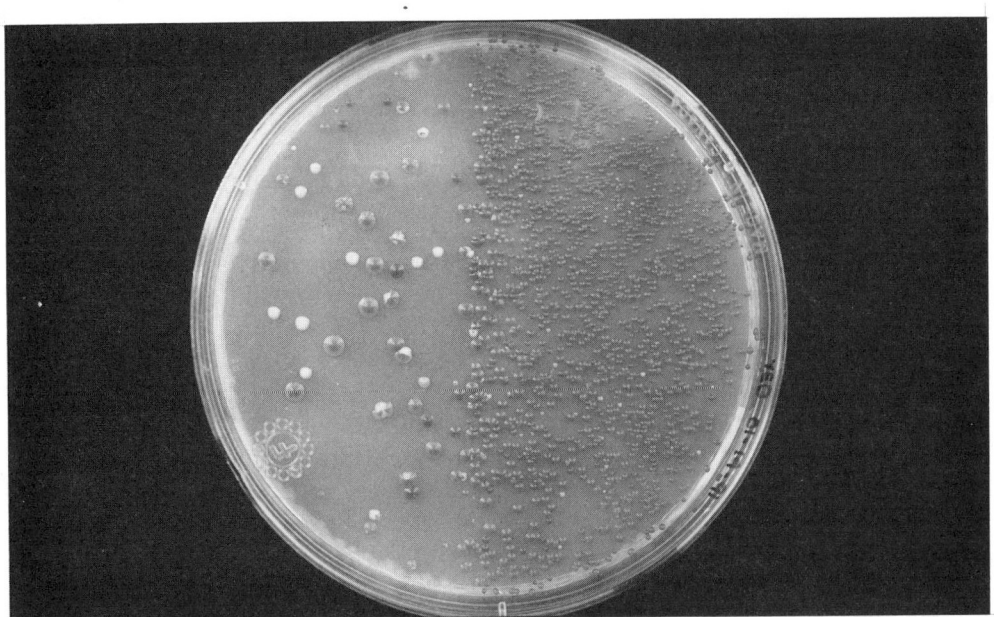

Figure 2. Yeast colonies on agar exposed and not exposed to UV radiation.

7. Invert the plate and incubate it at room temperature or in an incubator at 30° C.

8. After 2-3 days, count the number of colonies on both the experimental and control sides.

9. Record the data in the computer spreadsheet. Obtain class data and analyze the results. (If desired, the number of red and white/mutant colonies can be distinguished to gain an additional understanding of the effect of UV radiation on living cells.)

Results

The yeast on the portion of the petri dish exposed to UV radiation had fewer colonies than the nonexposed yeast (Figure 2). Table 1 shows the average results for six classes of 25-30 students that performed this experiment. Some classes had higher totals since the yeast dilutions varied with each class.

Conclusion

Students were asked to generate their own conclusions after observing the results of their experiment. Students have concluded that:

Yeast and UV Radiation

1. UV radiation kills yeast cells.
2. Controls are treated exactly like the experimental group except they are not treated with UV radiation.
3. Controls are necessary to analyze experiments.
4. A large sample size has a statistically better chance to give more accurate results than a small sample size.

Students quickly grasped the concepts of the scientific method. They learned to write a hypothesis predicting the outcome of an experiment and were able to analyze the class data and draw conclusions.

The term "control" was no longer a term they memorized out of their textbook. The control and the experimental treatment were readily apparent. We observed more critical thinking from our students as they began to analyze the experimental design of other labs.

This experiment is also an introduction to the concept of sterile technique.

Students noted that some plates were contaminated while others were not. This observation led to a student-generated discussion of how the plates should be handled to prevent contamination. In the microbiology unit, we found that students had gained considerable confidence in their use of sterile techniques.

After viewing the results of this experiment, students started to ask the following questions: What would happen if a longer or shorter UV exposure time was used? Could the irradiated yeast be protected with sunscreen? As students began to see extensions of this lab, they were asked to design their own experiments testing their new hypotheses.

Acknowledgment

This work was supported in part by grants DPE-8319148 and DTE-8954638 from the National Science Foundation under the direction of Tom Manney and Monta Manney. Their help was greatly appreciated.

References

American Cancer Society. (1992). *Cancer facts & figures-1992*. American Cancer Society, Pamphlet No. 92-425M.

Manney, T.R. & Manney M.L. (1990). *Handbook for using yeast to teach genetics*. Manhattan, KS: Kansas State University.

Yager, R.E. (1988). Never playing the game. *The Science Teacher, 55*(6), 77.

pH and Microbial Growth

John E. Lennox, The Pennsylvania State University, Altoona, Pennsylvania
Mary J. Kuchera

Most microbiology textbooks give considerable emphasis to the effect of pH upon microbial growth. Laboratory manuals also usually include exercises intended to illustrate the principles involved. In general, media adjusted to one of several pH values are prepared and students are instructed to inoculate these media with pure cultures of assigned organisms. The major differences in these exercises are the types of media used-- solid or liquid, and the choice of organisms (Atlas et al 1984; Benson 1985; Miller 1976; Seeley & VanDenmark 1981; Wistreich & Lechtman 1984).

These exercises do work and illustrate the lesson intended. Nevertheless, we know teachers who have encountered certain problems that tend to make the demonstration of pH effect less clear and therefore pedagogically less useful than, for example, the exercises demonstrating the effects of temperature or osmotic pressure.

Most of the problems encountered have to do with preparation of the medium. Some media, nutrient broth for example, produce a fine precipitate when adjusted to low pH. Dispensed into culture tubes, this precipitate is difficult to distinguish from bacterial growth. This precipitate can be avoided by filter sterilizing the low pH media, but this is more time-consuming than autoclaving and some instructional laboratories will not have the necessary equipment.

A somewhat different problem is encountered at low pH in solid medium preparation. If an agar containing medium adjusted to low pH is autoclaved, the polysaccharide is hydrolyzed and the medium will fail to solidify. To avoid this, the medium is autoclaved, allowed to cool somewhat, and then adjusted to the desired pH with sterile hydrochloric acid.

These minor difficulties in medium preparation, combined with difficulties in interpreting results, led us to look for an alternative demonstration of pH effect. The one described here is useful and can be set up in graduated cylinders as a demonstration for an entire class, or in culture tubes for individual student use.

We begin by autoclaving five half-liter graduate cylinders capped with aluminium foil and five (0.5 liter) flasks of nutrient medium. We have successfully used MRVP medium and nutrient broth supplemented with 0.5-1% dextrose. Using hydrochloric acid and sodium hydroxide the flasks of media are adjusted to the desired pH values (e.g., pH 3, 5, 7, 9 and 11). Each medium is poured into an appropriately labeled, sterile, graduated cylinder. The cylinders are placed uncovered in a spot where they can be conveniently observed over a period of time (Figure 1). As expected, the low pH cylinders will contain a fine precipitate, but this will usually settle within a few hours and will not subsequently interfere with the demonstration.

The observations can be made by eye if the cylinders are well lighted and backed up by

pH and Microbial Growth

a white card. Black lines (made with electrical tape) on the white card make it easier to estimate the degree of turbidity as microbial growth occurs. You can be precise by withdrawing samples and reading them in a colorimeter if one is available (Figure 2). If you choose to use a spectrophotometer, separate blanks for each of the several media will be necessary. Do not assume that media at different pH values will have the same absorbance.

The pH can be measured at intervals using a pH meter, but a pH test strip with multiple dyes capable of registering a broad pH range will also work well and has the advantage that the students can make the measurements rapidly themselves (Figure 3).

Finally, the samples withdrawn can be used to make stained slides for microscopic observation. Simple stains (crystal violet, methylene blue), Gram stain, spore stain and lactophenol cotton blue stain (for molds) are all useful in determining the general morphological types of organisms present.

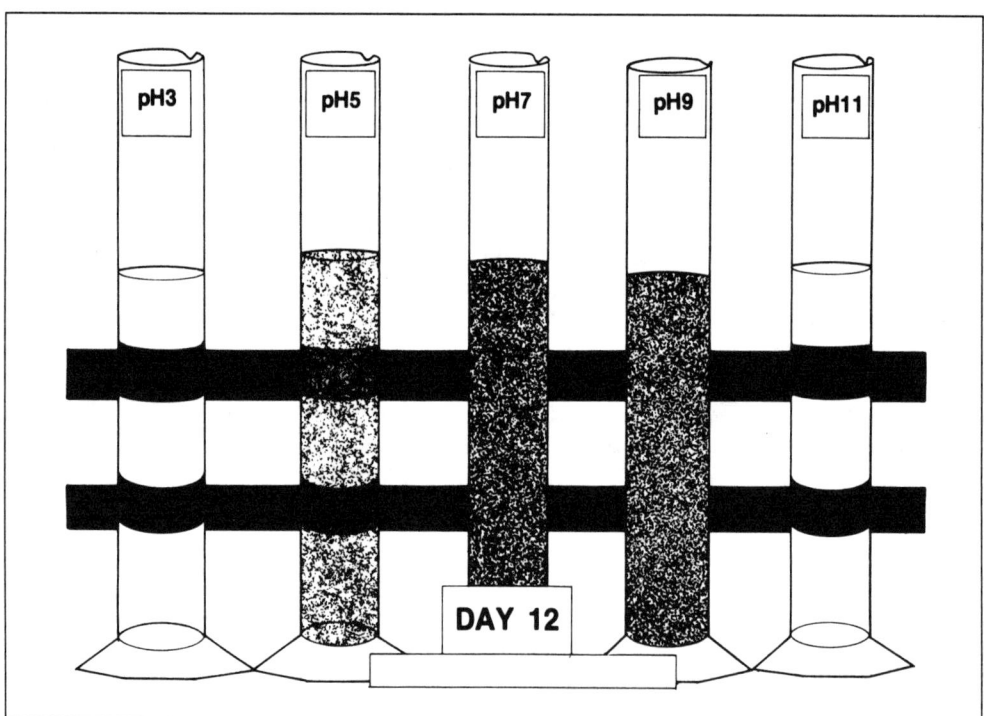

Figure 1. Five sterile half-liter graduate cylinders containing glucose nutrient broth adjusted to the indicated pH values are exposed to the air. Measurements of turbidity, pH changes and the microscopic examination of stained slides can be used to monitor growth of microorganisms under these different conditions. This represents the appearance of one experiment after 12 days of exposure.

Results and Discussion

pH and Microbial Growth

This demonstration, unlike most others, does not employ pure cultures but rather utilizes those airborne organisms that settle fortuitously into the cylinders. This being the case, you cannot predict with precision the sequence of events that will follow. Typically, the first change observed is in the pH 7 cylinder (Figure 2). Within a few days this medium will become turbid. This turbidity will most likely be associated with a decline in pH as heterotrophic organisms oxidize the dextrose to a variety of acidic substances (lactic acid, acetic acid, formic acid, etc.).

After several days of incubation, turbidity appears in the pH 9 and pH 5 containers. As Figure 3 indicates, these changes in turbidity may or may not be associated with changes in the pH of the medium. The students may also observe that the degree of turbidity in the several containers is not equal even at maximal growth (Figure 2, pH 5). Many students are surprised to see that in some cases the pH will rise after an initial drop. The initial drop is caused by the production of organic acids by fermentation of the glucose present. Later these same organisms, or perhaps successive waves of oxidative organisms, convert these organic acids to other compounds that do not contribute to the hydrogen ion concentration and the pH rises once again. Certain bacteria and fungi also produce alkaline metabolites, resulting in an increase in the pH.

The time course of the demonstration varies greatly between replicates reflecting differences in room temperature and available inocula. In any case, the sequence in which the cylinders exhibit growth is quite predictable. The pH 5 and 9 cylinders follow the pH 7 cylinder, and those at pH 3 and 11 are last to exhibit growth. The organisms that invade the cylinders at higher pH (7, 9 and 11) tend to be bacteria, although some surface growth of molds tends to form after a while in the pH 7 cylinder. By this time, of course, the pH of that cylinder is much lower than its label indicates (Figure 3).

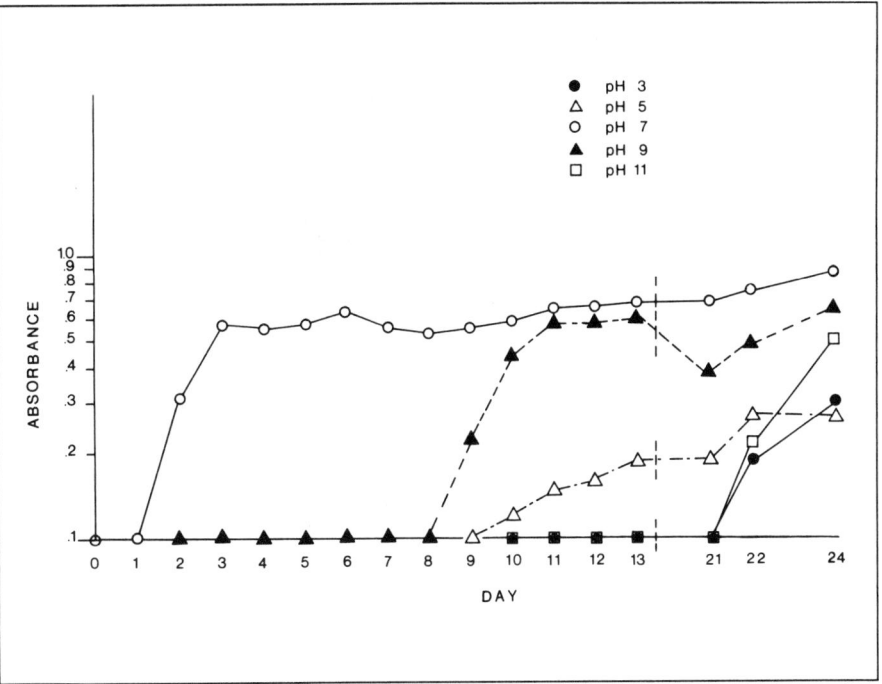

Figure 2. Absorbance of medium in each cylinder is plotted against time. Measurements were made with a Spectronic 20.

Biology Labs That Work: The Best of How-To-Do-Its

pH and Microbial Growth

In the cylinder at pH 5, molds and yeasts predominate, although bacteria such as the lactobacilli, are also found in most instances. It is not unusual for the pH 3 cylinder to remain without obvious growth until the end of the exercise; but in most cases, a few spherical fungal colonies can be seen growing at the bottom or on the surface of the cylinder. Teachers can use this observation as a point of departure to discuss why acidified and fermented foods have relatively long shelf lives. In some foods—pickles, beets, some meats and fish, etc.—acetic acid (vinegar) is added to lower the pH. In other products the natural fermentative activities of microorganisms acting on carbohydrates in the food result in a lowering of the pH and consequent increase in the shelf life. Sauerkraut, yogurt, buttermilk and certain cheeses are examples of the latter.

One could in theory prolong shelf life by raising the pH of foods, but alkaline substances with their bitter taste are not as pleasing to the palate as are products with low pH and a tart taste.

It may be instructive to ask students to list man-made or natural environmental niches that approximate the several conditions examined in this demonstration. Students might be provided with pH test strips and asked to construct a table similar to that in Figure 4.

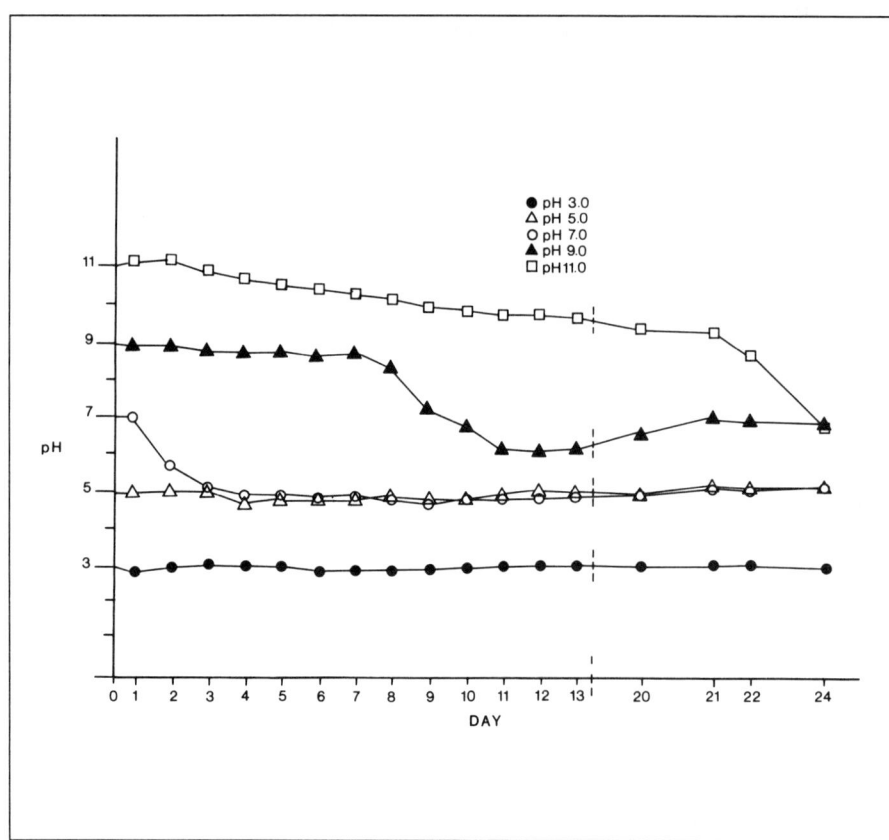

Figure 3. Changes in the pH of the medium are recorded at intervals during the experiment. In this exercise, a pH meter was used, but pH paper indicator strips could also be used.

We see no reason why this exercise could not be conducted in test tubes, giving each student a set of five tubes at different pH values. The sets could be placed by the students in different locations so they could compare the similarities and differences in growth patterns and organisms present.

We are confident that others will find many variations on this demonstration. That microorganisms have adapted to such diverse habitats as alkali soils (pH 9) and acid-mine damaged streams (pH 2) can be an astonishing observation. We hope that instructors of microbiology will find this demonstration useful.

pH and Microbial Growth

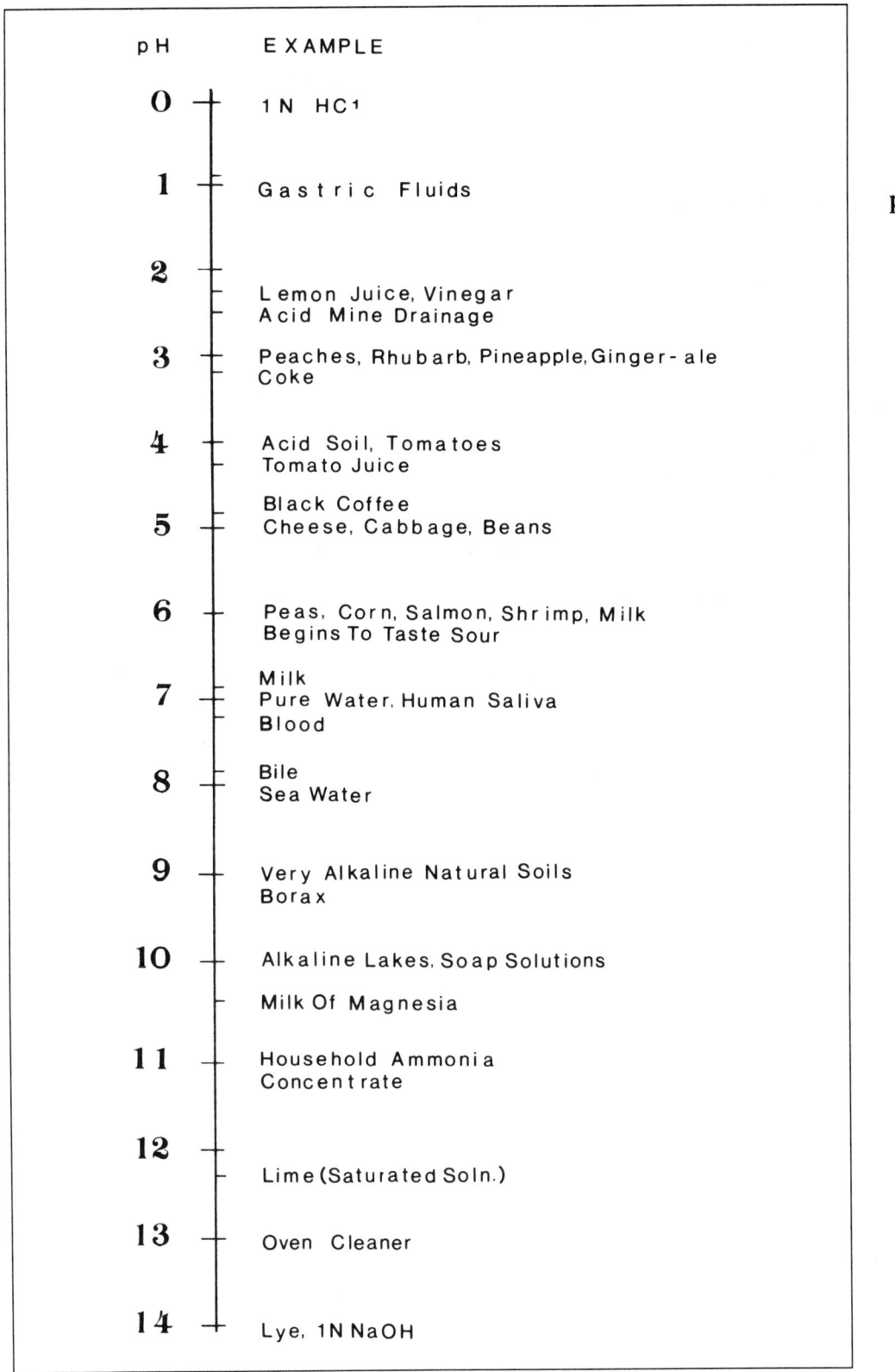

Figure 4. pH values of some common materials. pH values collected from a variety of contemporary textbooks.

pH and Microbial Growth

Acknowledgments

We thank Fran Laughlin and Susan Perkins for their assistance in preparing the graphic material for this paper.

References

Atlas, R.M., Brown, A.E., Dobra, K.W. & Miller, L. (1984). *Experimental microbiology: Fundamentals and applications.* New York: Macmillan Publishing Company.

Benson, H.J. (1985). *Microbiological applications: A laboratory manual in general microbiology, complete version,* 4th ed. Dubuque, IA: Wm. C. Brown Company Publishers.

Miller, A.I. (1976). *Microbiological laboratory techniques.* Lexington, MA: D.C. Heath and Company.

Seeley, H.W., Jr. & VanDenmark, P.J. (1981). *Microbes in action: A laboratory manual of microbiology,* 3rd ed. San Francisco, CA: W.H. Freeman and Company.

Wistreich, G.A. & Lechtman, M.D. (1984). *Laboratory exercises in microbiology.* 5th ed. New York: Macmillan Publishing Company.

Plants

Accurately Measuring Transpiration

David R. Hershey, Prince George's Community College, Largo, MD

Culp (1988) described transpiration measurements using a cut sunflower stem in a 100 ml graduate cylinder. This method was used by Sachs (1887) over a century ago (Figure 1). However, the drop in water level, as measured by cylinder graduations, will overestimate transpiration because the stem occupies a portion of the cylinder volume. For example, a 0.5-inch diameter stem in a 1-inch diameter graduate cylinder occupies 25% of the cylinder volume, causing transpiration to be overestimated by 33%.

A more accurate method of determining transpiration is to equate transpiration, in ml, with the weight loss, in grams, of the cylinder-stem apparatus, since the density of water is 1 g per ml.

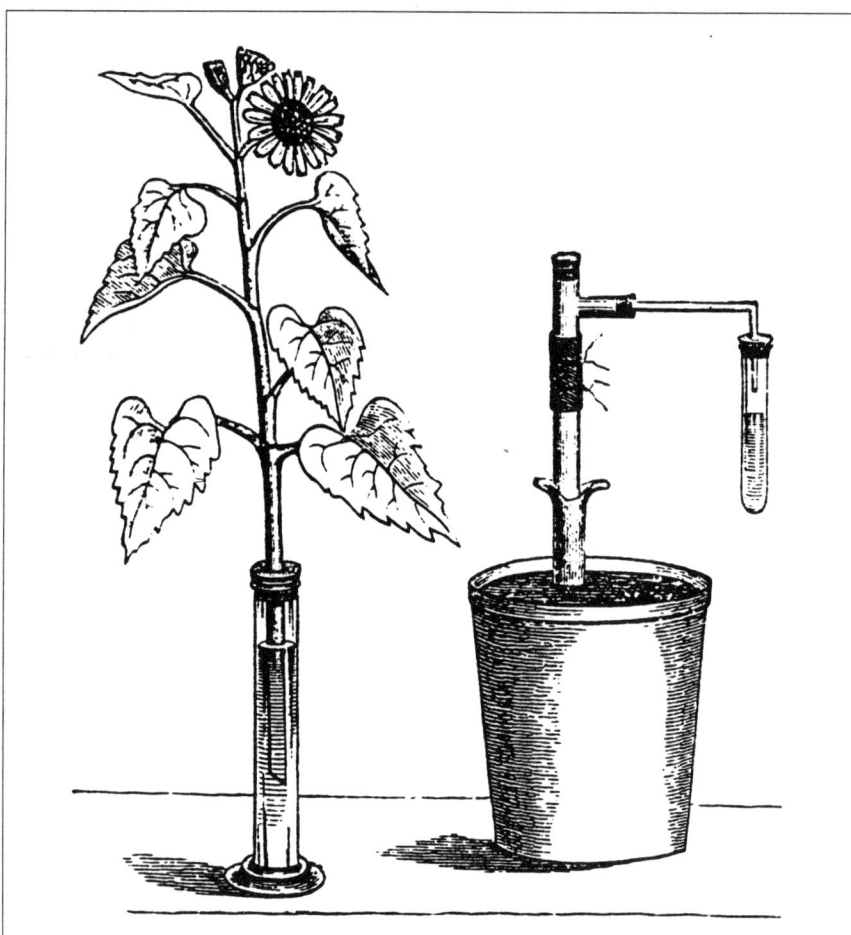

Alternately, a correction factor for stem volumes can be determined by first filling the graduate cylinder containing the stem to the 100 ml mark, then removing the stem and noting the water level in ml. This value represents the percentage of water level drop due to transpiration. For example, removing the 0.5-inch stem from the 1-inch diameter cylinder will cause the water level to fall from 100 to 75 ml. This means that only 75% of the water level drop will be due to transpiration. Therefore, if the water level drops 40 ml, the actual transpiration is only 0.75 x 40 = 30 ml.

References

Culp, M. (1988). An easy method to demonstrate transpiration. *The American Biology Teacher*, 50(1), 46-47.

Sachs, J.V. (1887). Lectures on the physiology of plants. London: Oxford. [Translated by H.M. Ward].

Figure 1. A cut sunflower stem placed in a graduate cylinder to measure transpiration. From Sachs (1887).

The Influence of pH on the Color of Anthocyanins and Betalains

Randy Moore, University of Akron, Akron, Ohio
Darrell S. Vodopich, Baylor University, Waco, Texas

Anthocyanins are a group of water-soluble pigments called flavonoids that are widespread among angiosperms, usually absent in liverworts and algae, and rare in mosses and gymnosperms. More than one anthocyanin often occurs in the same flower or organ. These pigments can impart violet, blue, purple, red and scarlet colors to flowers, stems, fruits and leaves.

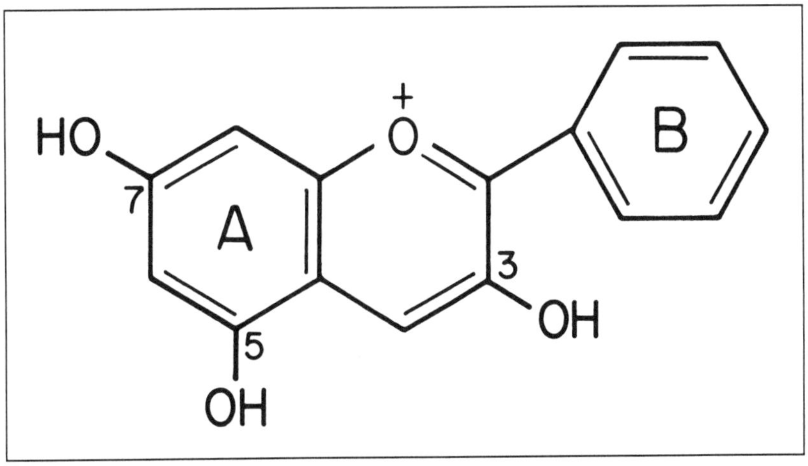

Anthocyanins are typically present as glycosides of one or two sugar molecules attached to anthocyanidin (Figure 1), which is also colored. The presence of sugars on the molecule increases its solubility and allows anthocyanins to accumulate in the aqueous sap of the central vacuole of plant cells. Hydroxylation or methylation of ring B of anthocyanidin (Figure 1) produces anthocyanins of different colors (Figure 2).

Figure 1. Structure of anthocyanidin. Position 3 of ring A is always glycosated with glucose galactose, rhamnose, xylose-glucose, rhamnose-glucose, or glucose-glucose. Position 5 is sometimes glycosylated--if so, it is glycosylated with glucose. Position 7 is almost never glycosylated--if so, it is glycosylated with glucose.

Anthocyanins probably attract pollinators to flowers and may also contribute to the spectacular color of autumn leaves. Yellow and orange carotenoids (e.g., in tomatoes), responsible for similar colors, are unrelated to anthocyanins.

The color of anthocyanins depends on several factors:

Associations with other compounds. For example, an association with phenolic compounds usually produces a blue color.

Substituent groups on the B ring of anthocyanidin. For example, the presence of methyl groups on this ring reddens the pigment.

pH of the cellular sap. Most anthocyanins are purple or blue at high pH and become reddish as the pH decreases. This pH-dependent change in color results from ionization of hydroxyl groups on the B ring, followed by electron shifts on the ring.

Betalains include red and yellow pigments called betacyanins and betaxanthins. Despite their similar colors, betalains are unrelated to anthocyanins and are found in only

Influence of pH on Colors

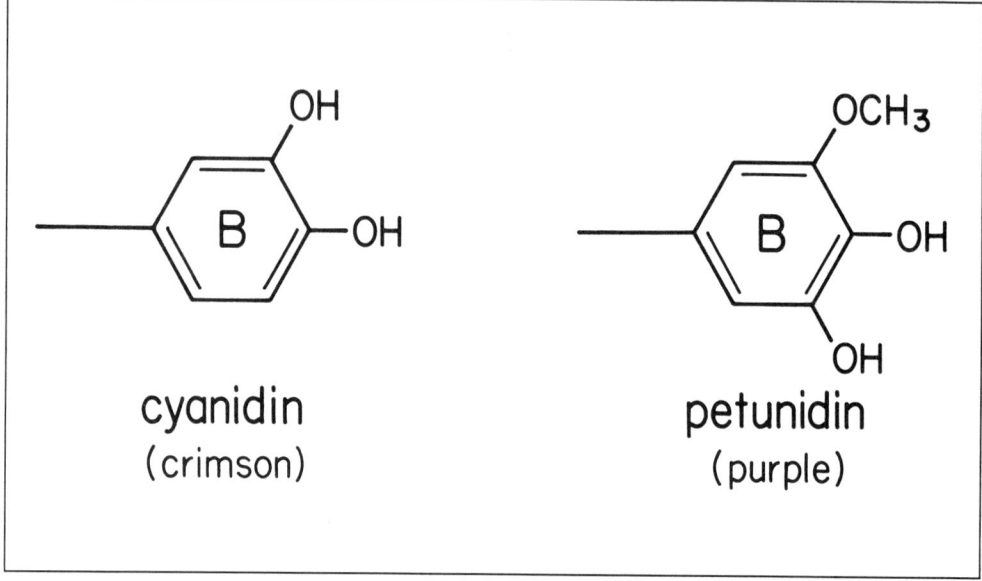

Figure 2. Cyanidin (first isolated from blue cornflower) is a hydrolated anthocyanidin, whereas petunidin (isolated from petunias) is a methylated anthocyanidin.

10 families of plants, all of which are members of the order Caryophyllales (Centrospermae). Indeed, betalains and anthocyanins do not occur in the same plant—individual plants cannot synthesize both of these pigments.

Betalains, like anthocyanins, are glycosides of sugars and colored groups. For example, betanin (the betalain in roots of red beets) is a glycoside of glucose and betanidin. Betanidin is red and has the structure shown in Figure 3. Like anthocyanins, betalains are water soluble and accumulate in the vacuole of plant cells.

Anthocyanins and betalains can be distinguished by their differing responses to changes in pH—betalains typically do not undergo extensive changes in color with pH as anthocyanins do. The following experiment is a "sure-fire," rapid and inexpensive means of demonstrating the influence of pH on the color of anthocyanins and betalains.

Objectives

This experiment demonstrates 1) the effect of pH on the color of anthocyanins and betalains, and 2) how a simple chemical test (i.e., changing the pH) can be used to distinguish between betalains and anthocyanins.

Materials

- Ten test tubes, 10-15 ml capacity
- Red beet roots, 20 g
- Red cabbage, 20 g
- Balance to weigh tissue

- Blender
- Filter paper
- Water aspirator
- 0.1 N HCl, 20 ml
- 0.1 N KOH, 20 ml
- 1 N HCl, 30 ml
- Buchner funnel
- Sidearm flask, 500 ml capacity
- Two beakers, 500 ml capacity, with labels
- 0.01 N KOH, 20 ml
- Crystalline KOH or NaOH, 20 pellets
- Test-tube rack

Influence of pH on Colors

Procedure

1. Separately homogenize 20 g of red cabbage and red beet roots in 400 ml of water in a blender.

2. Filter each homogenate through filter paper. This filtration is facilitated by using a buchner funnel in a sidearm flask attached to an aspirator.

3. Discard the filter and solid material.

4. Place the two extracts into separate, labeled flasks.

5. Add 5 ml of cabbage extract containing anthocyanin to each of 5 test tubes numbered 1-5.

6. Add 5 ml of beet extract containing betalains to each of 5 test tubes numbered 6-10.

7. Perform the following treatments:
   Tubes 1 & 6: None (untreated controls).
   Tubes 2 & 7: Add 1.0 ml of 0.1 N HCl, mix, and note any color change.
   Tubes 3 & 8: Add 0.5 ml of 0.01 N HCl, mix, and note any color change (the cabbage extract will become violet, or, if too basic, blue).
   Tubes 4 & 9: Add 1.0 ml of 0.1 N KOH, mix, and note any color change.
   Tubes 5 & 10: Add one pellet of KOH or NaOH, mix, and note any color change (the anthocyanin will become yellow).

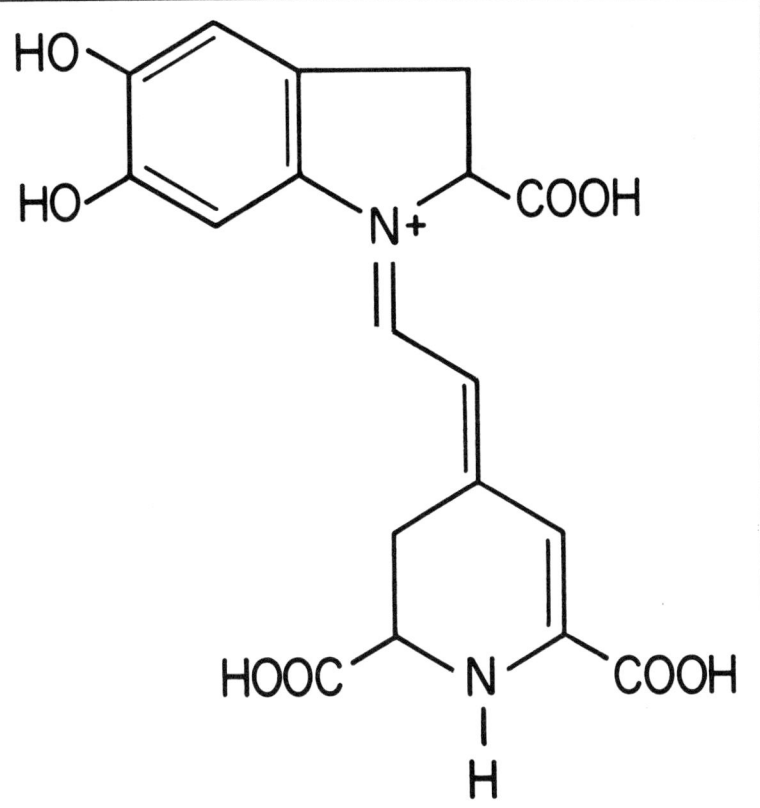

Figure 3. Structure of betanidin.

Influence of pH on Colors

8. These pH-dependent changes in pigment color are fully reversible. This can be demonstrated by adding drops of 0.1 N HCl to Tubes 3, 4, 8 and 9, or 1 N HCl for Tubes 5 & 10. Observe the changes in color.

Questions for Students:

Do betalains and anthocyanins show the same changes in color in response to differing pH values? Why or why not?

Can differences in pH account for all blue or red colors of flowers containing anthocyanins? Why or why not?

References

Goodwin, T.W. (Ed.) (1976). *Chemistry and biochemistry of plant pigments*. (2nd ed.). Vol. 1. London: Academic Press.

Plant Eco-Physiology: Experiments on Crassulacean Acid Metabolism Using Minimal Equipment

Douglas J.C. Friend, University of Hawaii, Honolulu, Hawaii

Crassulacean Acid Metabolism (CAM) provides opportunities for integrated studies in plant biology ranging from the ecological through the taxonomic, morphological, anatomical, physiological and biochemical. Simple supplies such as pH paper, plastic syringes and a solution of NaOH are sufficient for quite complex investigations, making the phenomenon of CAM ideal for class or individual investigations in the laboratory or field. In addition, experiments can be performed in advance and the plant material frozen and stored until a time convenient for class analysis.

CAM is a complex eco-physiological plant adaptation to the problems of growth in an arid environment. The normal diurnal pattern of stomatal movement is reversed. Stomata are open at night, allowing dark CO_2 uptake and storage of malic acid. During the day, stomata are closed, greatly restricting water loss. Photosynthesis still occurs because CO_2 is supplied inside the leaf by decarboxylation of the stored malic acid. A summary of the main features of CAM plants follows (Gibson 1982; Kluge & Ting 1982; Nobel 1988; Osmond 1978; Osmond & Holtum 1981; Ting & Gibbs 1982).

Features of CAM Plants

Taxonomy. The accumulation of acid in plant leaves in the Crassulaceae led to the naming of the whole syndrome as Crassulacean Acid Metabolism. The CAM habit has apparently evolved independently in several other families such as the Cactaceae, Portulacaceae, Orchidaceae, Agavaceae and Bromeliaceae. There are even CAM ferns. Pineapple and sisal are two commercial crop plants with CAM.

Ecology. CAM plants are often present in extreme desert environments. Physiological water stress imposed by saline soil conditions may also induce CAM characteristics called facultative CAM plants. Low night temperatures and short day lengths often lead to increased acid accumulation.

Morphology. Leaves are often thick and fleshy (succulents). Some plant groups such as the cacti store water in swollen stems, the leaves being very reduced.

Anatomy. Leaves are adapted to a low rate of water loss (low transpiration) by the following characteristics: thick cuticle; low stomatal density; thick chlorenchyma layer directly below the epidermis, often not divided into palisade and spongy mesophyll layers; presence of large vacuoles in chlorenchyma cells and presence of large, thin-walled cells of water-storing tissue beneath the chlorenchyma (Gibson 1982).

Physiology. Stomata close during the day, restricting water loss. At night the stomata open and CO_2 is taken up from the air. Water loss through open stomata is lower at night than it would be during the day because of the lower temperature and higher humidity. The uptake of CO_2 is thus separated in time from its reduction to sugars and starch during photosynthesis in daylight.

Experiments on CAM

Biochemistry. The enzyme phosphoenolpyruvate carboxylase (PEP carboxylase) is present in the cytoplasm of mesophyll cells and fixes CO_2 from the air at night. The 3-carbon compound phosphoenolpyruvate (PEP) is the acceptor for CO_2 and is formed in the leaf by glycolysis from stored starch or sugars. CO_2 fixation produces the 4-carbon compound malic acid which is stored in the mesophyll vacuoles. This nighttime accumulation of malic acid gives a low pH and sour taste to the cell sap of CAM plants, hence the term acid metabolism.

During the day, the malic acid diffuses from the vacuoles to the mesophyll cytoplasm and/or mitochondria where it is decarboxylated to produce gaseous CO_2 and pyruvic acid. A high concentration of CO_2 is held inside the leaf because the stomata are closed. The CO_2 is then refixed into carbohydrate by the usual process of photosynthesis in the chloroplasts, through the action of the enzyme ribulose biphosphate carboxylase oxygenase (RUBISCO). The NADPH and ATP needed to reduce CO_2 to carbohydrate are formed in the chloroplasts as a consequence of light-activated electron transport during photosynthesis. The pyruvate remaining after decarboxylation of the malic acid is incorporated into PEP and used to form sugars and starch. PEP can be regenerated the following night and used again to fix more CO_2 into malic acid (Figure 1). There are

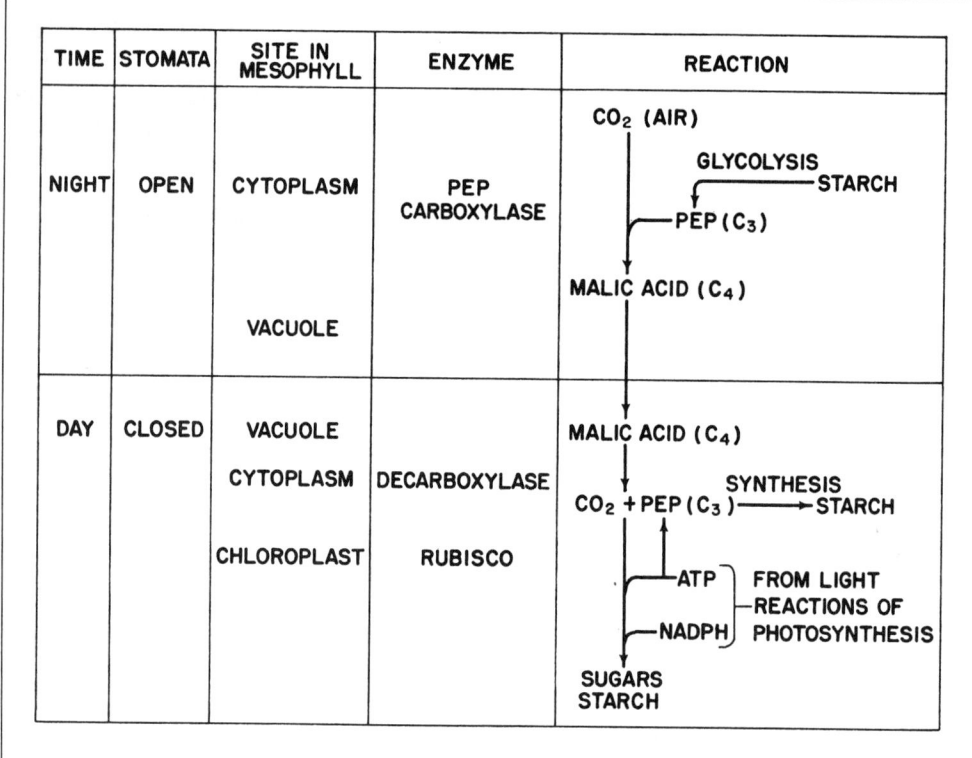

Figure 1. Generalized pathway of CO_2 uptake in CAM.

several variants on this basic pattern of CAM. Aspartic acid may be stored instead of malic acid, and there are variations on how PEP is reformed.

Experiments on CAM

Measurement of CAM

Squeeze the sap from leaves and measure the pH with indicator paper or a pH meter and/or titrate the acid with NaOH.

Plant material. Any succulent plant may be used. Commonly available plants include the weed *Portulaca oleracea* and horticultural varieties of the garden plants *Sedum spectabile, Kalanchoe blossfeldiana, Mesembryanthemum, Sansevieria, Aloe, Opuntias* and other common cacti. Depending on climate, these plants may be available as garden or houseplants. Fleshy *Euphorbias* are best avoided as some of these contain irritating latex. In Hawaii I have used *Kalanchoe daigremontiana*, a common domestic plant that has become wild in many places.

Treatment of leaves. Temperature or light treatments may be given to isolated leaves as well as to whole plants. Place detached leaves with the lower epidermis down on a tray of damp vermiculite or paper towels or individually in small open plastic bags. Eight samples of about 10 g of leaves should be used for each treatment at each time. Select leaves of similar morphology and position on the plant. After treatment, place leaves in labeled plastic bags and seal and store them in a freezer. This will not only halt further metabolism but also cause rupture of membranes during subsequent thawing, so the contents of the vacuoles are adequately sampled. Frozen leaves may be stored for several months with no change in acid levels, as determined previously.

Extraction of sap. Thaw the leaves and extract the cell sap by rolling a dowel or bottle over the leaf, making sure that the sap collects at the base of the plastic bag. Alternatively, use a small vise or garlic press.

Measurement of pH. Use narrow range pH papers, dipping them in the sap. Papers with a range of about 3 to 5.5 are suitable (e.g., Hydrion short range paper #425 pH 3.0-5.5, #430 pH 3.4-4.8). Alternatively, use a pH meter.

Measurement of acid concentration.
1. Prepare a solution of 0.1N NaOH (4 g/liter) and fill a 5 or 10 ml plastic disposable syringe (minus needle).

2. Use a 1 ml plastic syringe (minus needle) to pipet 1 ml of juice from the plastic bag into a small paper portion cup (disposable) or beaker. Add three drops of a phenolphthalein indicator (0.05g in 50 ml ethanol + 50 ml water).

3. Adjust the NaOH syringe to the zero mark and slowly add solution to the sap in the cup. Continue adding NaOH, swirling the cup all the time, until the solution reaches a red-brown end point (about pH 9) that does not change with continued swirling.

Biology Labs That Work: The Best of How-To-Do-Its

Experiments on CAM

Record the ml of NaOH used, remembering that the syringe usually records the amount left rather than the amount delivered. The end point will differ in color depending on the plant material. Flavenoids present in the sap will turn yellow as the pH increases near the end point. Anthocyanin pigments, initially pink, will become blue-green as the end point is approached. Students should perform some titrations without phenolphthalein to be aware of these color changes and be able to explain them. The pigments naturally present in the sap might even be used as an indicator. Standardize the color of the end point for every individual and have each student titrate at least one replicate of each treatment. Class values will then average out different evaluations of the end point. The molar concentration of diprotic acids is obtained by dividing the ml of 0.1 N NaOH by 2.

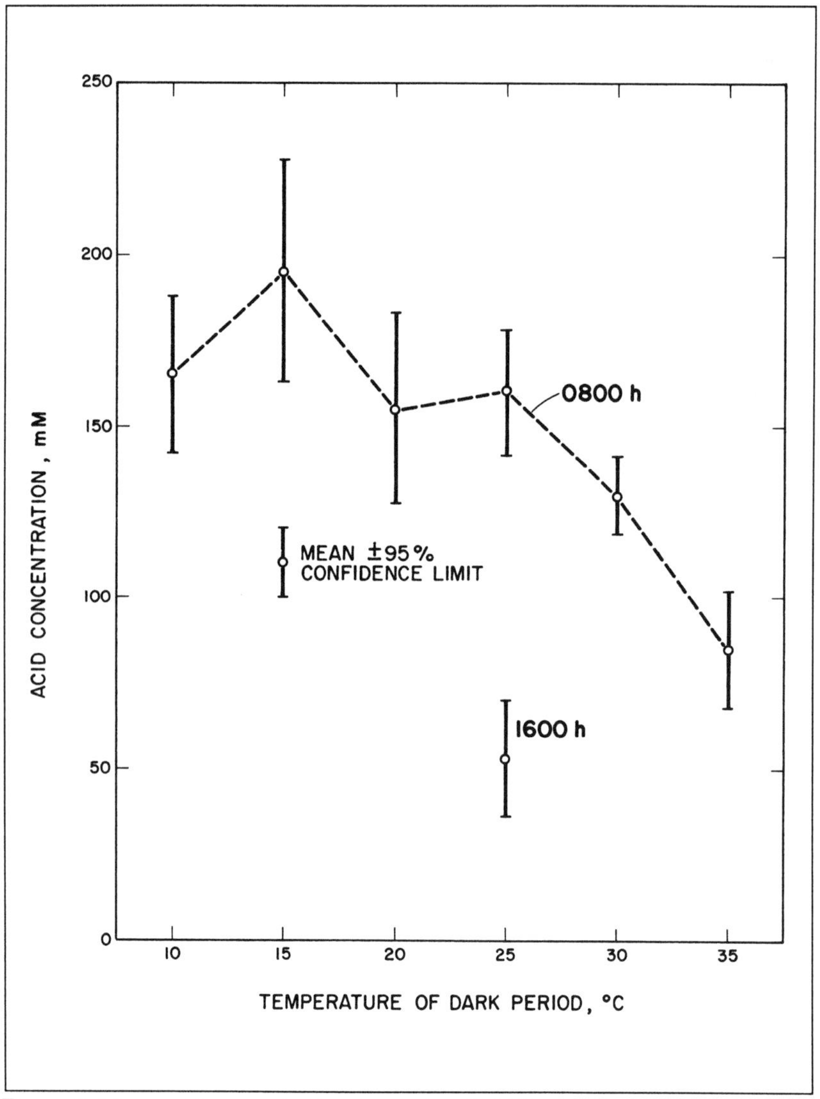

Statistical treatment. Statistical significance is quickly measured by using range statistics, a rapid way of obtaining a confidence limit for the mean of a small number of observations (less than 10). Comparing two means plus or minus their confidence limits is equivalent to a t-test. (For further details see Dean and Dixon (1954) and Table 1.) Use of confidence limits gives students a good feel for the necessity of adequate replication

Figure 2. Mean acid concentration (upper) and pH (lower) during day and night in detached leaves of *Kalanchoe daigremontiana* maintained at 25° C in natural daylight (n=9).

Table 1. Confidence limit factors (P = 0.95) derived from range of replicates (from Dean & Dixon 1954).

Number of replications	Confidence limit factor
2	6.4
3	1.3
4	0.72
5	0.51
6	0.40
7	0.33
8	0.29
9	0.26
10	0.23

and an appreciation of biological variability. An example for the calculation of the confidence limits of two points from Figure 2 is given in Table 2.

Experiments and Investigations on CAM

The following list provides suggestions for experiments on the physiological aspects of CAM as these are probably the most easily adapted to laboratory periods.

1. **Comparison of CAM and non-CAM plants.** Select a nonsucculent plant such as spinach and a succulent plant that is suspected of having CAM. At the end of the day, label and freeze leaf samples from both types of plants in separate plastic bags. The following morning, as early as possible, take similar samples and freeze them. Determine the acid content of the expressed sap from thawed leaves. Overnight accumulation of acid is characteristic of CAM plants.

2. **Time course in pH and acid accumulation in a CAM plant.** Select a succulent plant in which you know that CAM takes place. At the beginning of the day, label and freeze all samples as before. If detached leaves are used, place them on a tray of damp paper or vermiculite or in separate unsealed plastic bags (the exact number will depend on how many times samples are taken). Place the leaves in daylight or high intensity artificial light for eight hours. If daylight is used, prevent overheating in direct sunlight by placing plastic or glass dishes containing about 5 cm of water over the leaves to act as an infrared filter. Label and freeze leaf samples at two- to four- hour intervals. After eight hours, place leaves in darkness for 16 hours, sampling and freezing leaves at intervals as before. A second light period could be sampled similarly. For class use, thaw the leaves and give them to students for juice extraction, and acid and pH measurement. Results from a typical lab experiment are given in Figure 2.

Table 2. Calculation of confidence limits of means. Examples are taken from Figure 2.

	Acid concentration, mM	
Replicate #	1900 h	0300 h
1	35	120
2	45	110
3	30	125
4	40	100
5	75	75
6	60	140
7	35	105
8	70	90
9	70	90
MEAN	51	106
RANGE	75 TO 30 = 45	140 TO 75 = 65
RANGE × CONFIDENCE LIMIT FACTOR		
	45 × 0.26* = 12	65 × 0.26 = 17

*Confidence limit factor from Table 1 for 9 replicates = 0.26
Do means plus or minus confidence limits (CL) overlap?

1900 h mean CL = 51 +/- 12 = 63 to 39
0300 h mean CL = 106 +/- 17 = 123 -89
There is no overlap so means do differ significantly.

Experiments on CAM

3. **Effect of CO_2 level during the dark period on acid accumulation.** Sample and freeze CAM leaves at the end of the day. Seal other samples in airtight plastic bags together with about 5 ml of soda-lime in a paper cup, taking care not to spill the soda-lime. This will provide a CO_2-free environment during the dark period. Sample other leaves and place them in open bags without soda-lime, as normal-air controls. Place leaves in darkness for 16 hours, and the following morning remove the soda-lime, label all bags, and freeze the leaves. Thaw the leaves later and titrate as previously. Did absence of CO_2 during the dark period affect the ability to accumulate acid? Results from a class experiment are given in Table 3.

4. **Effect of an extended period of darkness on acid accumulation.** At the end of the day, freeze leaves in plastic bags. Place experimental leaves in darkness. Remove leaf samples at two- to four-hour intervals, place in plastic bags, label and freeze. Continue taking samples in darkness until the following morning. Transfer half the remaining experimental leaves to light. Continue sampling leaves both in light and in dark until the following evening. Is there any evidence for a continued accumulation of acid when darkness is prolonged beyond the normal limit? Deacidification during the later stages of a long period of darkness would suggest the involvement of an endogenous rhythm in CAM. Results from a class experiment are given in Figure 3.

5. **Effect of temperature on acid accumulation during the dark period.** Freeze leaves at the beginning of the dark period. Place other leaves in darkness at a range of temperatures from about 10° - 35° C or whatever temperatures are available as long as they are recorded. Sample and freeze leaves from each temperature treatment at the end of the dark period. What is the optimal temperature for acid accumulation? Results from a class experiment are given in Figure 4. With plenty of plant material and a number of students involved, it would be interesting to carry out a time-series of sampling during the dark period at different temperatures to see whether the rate of acid accumulation is affected by temperature.

Table 3. CO_2 effects on acid accumulation and pH of detached leaves of *Kalanchoe daigremontiana* during a 16-hour dark period (n = 9).

	End of 8-hour day	End of 16-hour night $-CO_2$	End of 16-hour night $+CO_2$
Acid conc. mM	55 +/- 15	80 +/- 11	120 +/- 13
pH	4.6 +/- 0.05	4.1 +/- 0.09	3.6 +/- 0.14

Presence of CO_2 resulted in significant acid accumulation (non-overlapping confidence limits for means at end of 8-hour day and end of 16-hour night). Absence of CO_2 resulted in no significant acid accumulation (overlapping of confidence limits for mean at end of 8-hour day and 16-hour night).
*+/- confidence limit at P = 0.95 from range statistics.

6. **Effect of light intensity during the day on deacidification.** Freeze samples before and after a 16 hour period of darkness. Place experimental leaves under a range of light intensities, obtained either by using shade cloth and natural daylight or a range of intensities of fluorescent or incandescent light. Take samples from each light condition at the end of the day, or at two- to four-hour intervals if possible. Did the light intensity during the day affect deacidification?

As a variation, determine whether the light intensity given during the day affected the rate of subsequent dark acidification and intensity needed to prevent dark acidification. Use colored filters to determine the spectral regions most effective in preventing acidification.

7. **Transpiration of CAM plants.** Place leaves in small, labeled glass or plastic vials with the petioles dipping in water. Float a small amount of corn oil on the water to prevent evaporation and restrict water loss to the leaves. Weigh the vials plus leaves to an accuracy of about 0.05 g and note the weight and time on the label of each flask. Place leaves and flasks in light for about eight hours and reweigh, then transfer them to darkness for about 16 hours and reweigh. Record leaf areas by removing the leaves and placing them under a transparent grid of dots spaced 25 mm apart. Such a grid is easily made by dotting the corners and center of each 1-cm square on a piece of cm graph paper and making a transparent Xerox copy. Count the total number of dots that cover the leaf. Dividing the total number of overlapping dots by 4 gives the area in cm^2. Calculate rates of transpiration as g water lost per m^2 per hour.

The rate of transpiration can be correlated with stomatal opening during the light and dark periods by using a dye infiltration method. Make a solution of 0.01g liter crystal violet or other alcohol-soluble dye in absolute ethanol and place it in a tightly stoppered dropping bottle. Dilution of the absolute ethanol by atmospheric moisture will reduce the ability of the dye to penetrate open stomata. A qualitative measure of the opening of the stomata is obtained by placing a few drops of the dye on each side of the leaf lamina. After about 30 seconds, wash the leaf under running water to remove surface dye. Areas of the leaf where stomata were open will be visible to the naked eye by the penetration of the dye into the mesophyll air spaces.

CAM provides many opportunities for individual projects over a range of levels than can be matched with student interests. There are a number of reviews of CAM (Kluge & Ting 1982; Osmond 1978; Osmond & Holtum 1981) as well as several books at the general (Nobel 1988) and more advanced levels (Ting & Gibbs 1982). The integration of information from ecology, morphology, anatomy, physiology and biochemistry provides a useful review of important botanical concepts. The biochemistry of

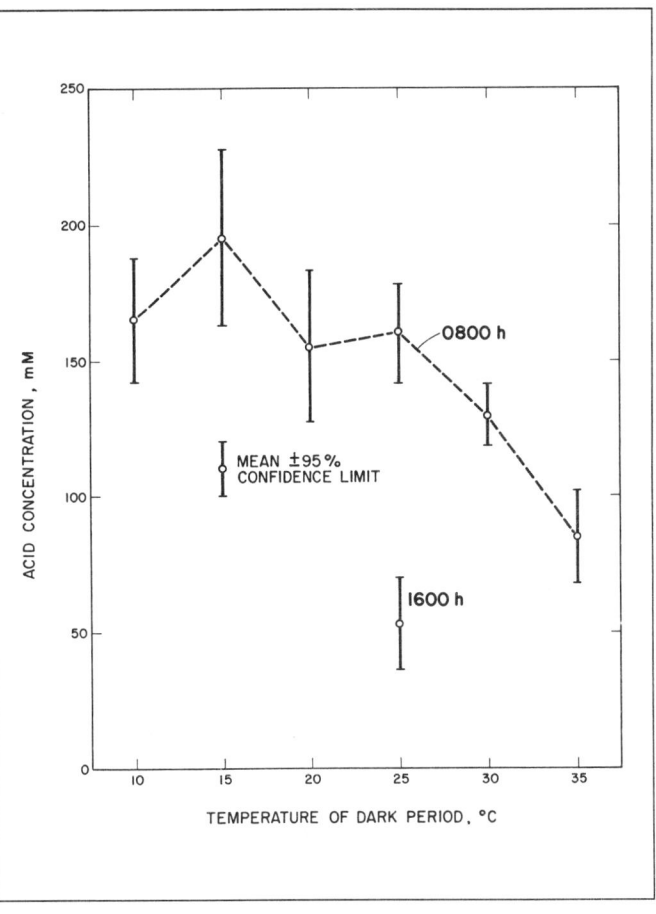

Figure 3. Dark inhibition of deacidification phase of detached leaves of *Kalanchoe daigremontiana*. Leaves maintained at 25°C in darkness or daylight (n=10).

Experiments on CAM

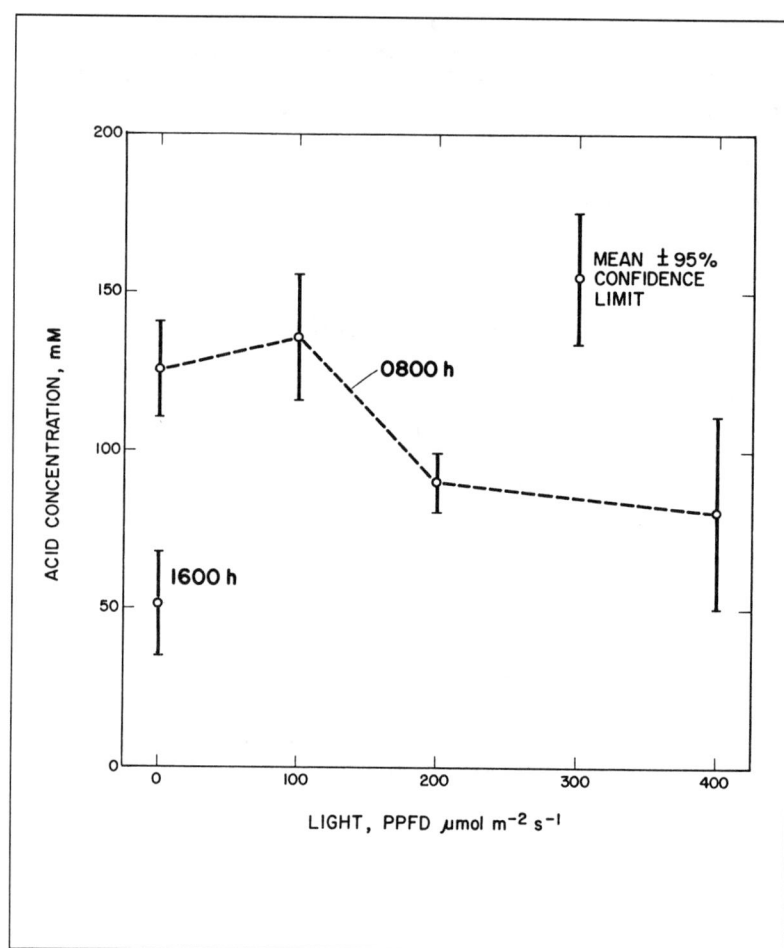

Figure 4. Night temperature effects on acid content of detached leaves of *Kalanchoe daigremontiana* (n=10).

CAM for instance is a variant on the C_4 metabolism found in a number of tropical grasses such as corn and sugar cane. Experiments on the biochemistry of CAM have not been included because of their complexity and the need for apparatus such as an HPLC for identification of dicarboxylic acids, a spectrophotometer for following enzyme activities, etc. If this equipment is available, a good starting point for references on procedures would be Ting & Gibbs (1982) and Nishida & Hayashi (1980).

Acknowledgments

Thanks are expressed to the students in my plant physiology labs at the University of Hawaii for the results in Figure 1 and Figure 3, and Mr. T. Nagata for preparation of the figures.

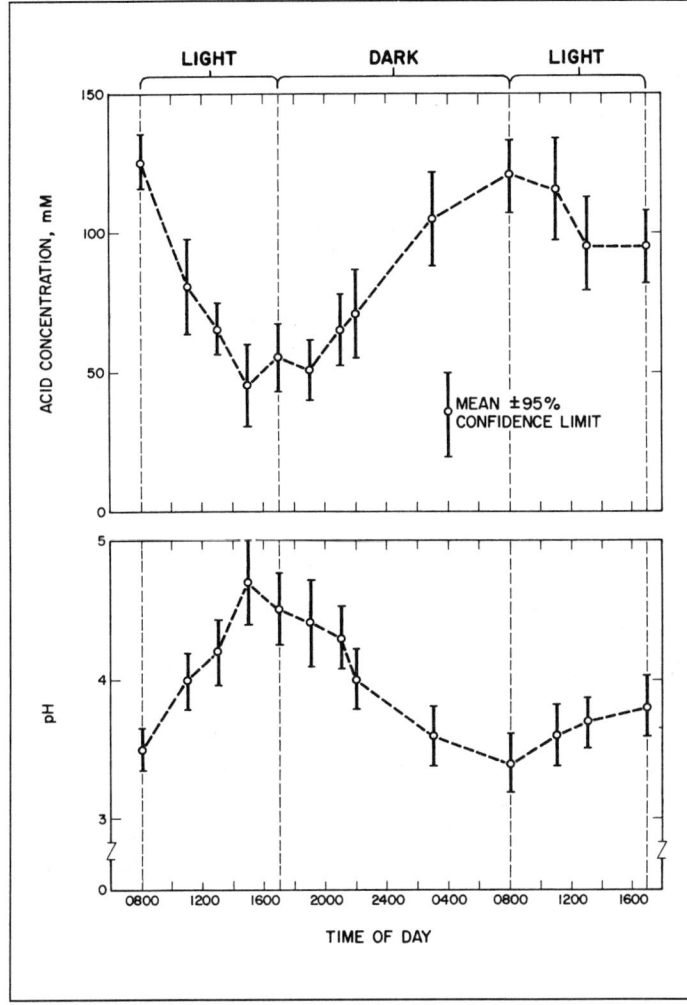

Figure 5. Light intensity effects (PPFD=photon flux density of photosynthetically active radiation between 400 and 700 nm) on nocturnal acidification of detached leaves of *Kalanchoe daigremontiana*. Leaves maintained at 25°C in darkness or at stated PPFD provided by white fluorescent lamps for 16 hours, after 8 hours natural daylight (n=10).

References

Dean, R.B. & Dixon, W.J. (1954). Simplified statistics for small numbers of observations. *Analytical Chemistry, 23*, 636-638.

Gibson, A.C. (1982). The anatomy of succulence. In I.P. Ting & M. Gibbs (Eds.), *Crassulacean acid metabolism* (pp. 1-17). Baltimore, MD: Waverly Press.

Kluge, M. & Ting, I.P. (1982). Crassulacean acid metabolism: Analysis of an ecological adaptation. *Ecological Studies, 30*.

Nishida, K. & Hayashi, Y. (1980). Inhibition of deacidification (loss of titratable acidity) by photosynthetic inhibitors in leaves of a CAM plant. *Plant Science Letters, 19*, 271-276.

Nobel, P. (1988). *Environmental biology of agaves and cacti*. Cambridge, MA: Cambridge University Press.

Osmond, C.B. (1978). Crassulacean acid metabolism, a curiosity in context. *Annual Review of Plant Physiology, 29*, 379-414.

Osmond, C.B. & Holtum, J.A.M. (1981). Crassulacean acid metabolism. In M.D. Hatch & N.K. Boardman (Eds.), *The biochemistry of plants* (Vol. 6, pp. 283-328). New York: Academic Press.

Ting, I.P. & Gibbs, M. (1982). *Crassulacean acid metabolism*. Baltimore, MD: Waverly Press.

Using Dandelion Flower Stalks for Gravitropic Studies

Paul E. Clifford, The Queen's University of Belfast, Belfast, Northern Ireland
Edwin L. Oxlade, Stranmillis College, Belfast, Northern Ireland

The development of tropic curvatures by plant organs in response to gravity is a subject of continuing debate among plant biologists. In recent years some basic observations and early experimental results have been re-examined. Consequently, some reviews of the subject (Firn & Digby 1980; Wilkins 1984) are more likely to question old ideas than was the case 20 or 30 years ago.

From Darwin's time to the present day, investigators of gravity-induced growth curvatures have most commonly used cereal coleopiles or seedling roots as experimental material. These are not easy materials, either to prepare or to work with; for teachers wishing to carry out practical work, there are alternatives.

Would you like to make plant gravitropism more than a theoretical study in your classroom without having to deal with materials such as roots or coleopiles? For several years we have used dandelion flower stalks both for our own research into the mechanism of gravitropism and as a convenient, manageable and effective material for classwork (Oxlade & Clifford 1981). This paper describes simple ways in which the gravitropic response of flower stalks may be investigated. None of the experiments requires more than the normally available materials of a school biology laboratory nor involves extensive periods of time.

The Material

The dandelion, *Taraxacum officinale*, Weber, is a common weed on both sides of the Atlantic. In early summer each plant produces large numbers of bright yellow inflorescences supported by long, straight, hollow stalks called scapes. These flower stalks, or scapes, are strongly negatively gravitropic up to the time the inflorescence opens. Thereafter they are weakly negatively gravitropic and may even be digravitropic during the stage when the inflorescence closes up and the yellow corolla tubes wither and fall off. Following this "closed head" stage there is a renewed surge of growth just prior to the formation of the fruiting head or clock, as it is sometimes called, when, again, the scape is strongly negatively gravitropic.

Because the gravitropic behavior of the scape varies depending on the stage of development of the inflorescence, the experiments described here are best carried out with scapes taken when the inflorescence is still in bud. At this stage the scapes are growing strongly and respond well to gravity.

Collecting the Scapes

Choose straight scapes at least 15 cm in length, cutting through them at the base with a sharp knife or razor blade. Handle the scapes as little as possible and make sure they are kept upright from the moment they are collected until they are used in experiments. Any deviation from the vertical at this stage could cause a subsequent unplanned gravitropic response. One convenient way to carry the scapes upright is in a measuring cylinder containing a small amount of water. If the scapes are kept standing in water they will not only last well but will continue to grow for several days.

Observing the Gravitropic Response

A scape's gravitropic response can be seen when it is moved from a vertical to a horizontal position and left to reorient itself by bending. Hold the scape while observing its tropic response by pushing its base into a water-filled glass tube slightly larger in diameter than the scape itself. To make suitable tubes, cut 15 cm lengths of glass tubing 6 mm in diameter and seal one end by melting the glass in a Bunsen flame.

Push each scape into a tube so that at least 10 cm of the scape is left free. To prevent any tendency for the material to rotate or move up and down, wedge a small piece of Plasticine at the open end of the tube. Again, it is important to keep the scapes in a vertical position at all times while fixing them in their tubes so as not to gravistimulate them prematurely.

Once the scapes are fixed in their tubes they can be decapitated by cutting through the scape just below the inflorescence bud. The gravitropic response will in no way be diminished by removing the inflorescence.

To gravistimulate (reorient to the horizontal position) the scapes, simply lay the tubes in a horizontal position at the edge of a bench top. Fix the tubes to the bench top with sticky tape or strips of Plasticine. Record the time when each scape is placed horizontally and watch what happens.

Figure 1. The dandelion scape is mounted in its water-filled glass tube, a 10-cm length marked and the inflorescence bud cut off. The tube is placed in a horizontal position at time 0.

Gravitropic Studies

Lag Time and Bending Rate

Gravistimulation is followed by a period during which no observable bending of the scape takes place. This is known as the lag time, reaction time or latent period. The lag time for each scape can be found by recording the time following gravistimulation at which you observe the first upward movement of the scape's cut tip. It will vary from scape to scape, but should average somewhere between 20 and 30 minutes. Investigation of the lag time alone and factors that might affect it could provide an interesting practical session, easily completed in less than one hour.

Once the lag time has passed, the scapes should be seen to bend progressively upwards at a rate that can be measured either as height of tip against time or tip angle against time. Rates of bending vary considerably. Perhaps the most important factor in determining this variability is the stage of development of the inflorescence when the scape was collected; other factors such as temperature will also play a part. Some scapes can bend at an impressively rapid rate, and the tip of a 10-cm length may reach the vertical stage in as little as two hours from the end of the lag time.

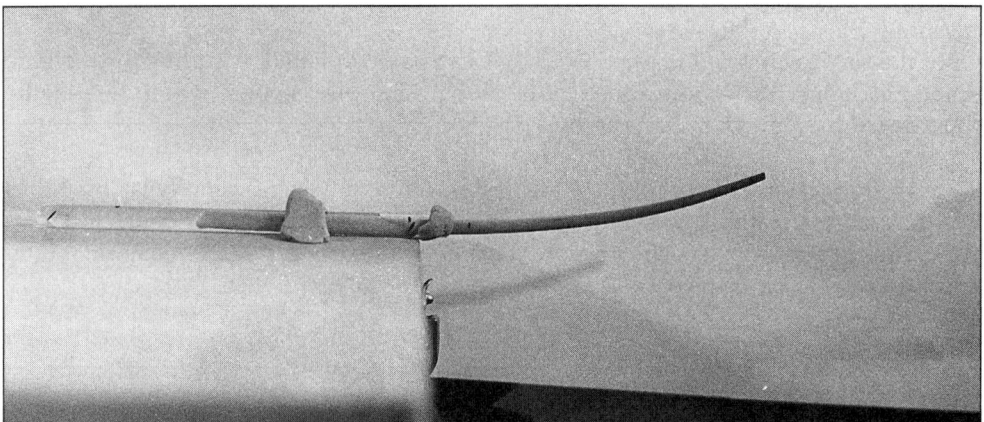

Figure 2. Forty-five minutes later the latent period has passed and the scape is starting to bend upward.

Autotropism

A little known feature of the gravitropic response is that an organ such as a flower stalk which bends following gravistimulation, will later straighten--a process that has been called autotropism. The dandelion scape shows this very well. Once a 90° bend has been achieved in a horizontally placed scape, the cut tip of the scape will be vertical and the 10-cm length of free scape will form a perfect quarter circle. Any further bending without compensatory changes would actually bring the tip away from the vertical again, which does not happen. Neither does growth of the scape stop when the tip just reaches vertical. What will be observed is that the tip remains vertical and regions below it straighten as other regions further below the tip continue to bend. Eventually the scape will be almost entirely vertical, with a tightly bent section where it is inserted in the glass tube.

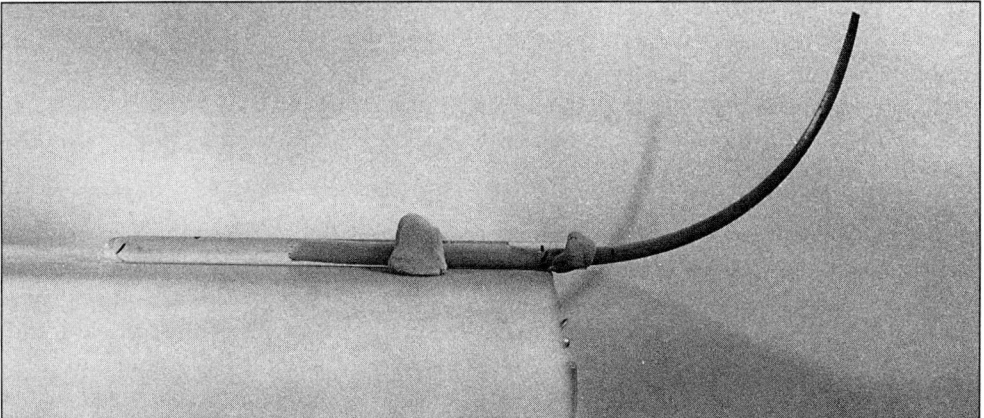

Figure 3. After three hours, the tip of the scape has almost reached a vertical position.

Growth Rate Changes

A tropic response is, by definition, a growth response. Bending of the dandelion scape is brought about by a difference in growth rates between the upper and lower sides of the organ. Upward bending occurs because the lower side grows faster than the upper side. Gravistimulation, in causing growth rate changes and upward bending, might affect only one side of the scape. Either the upper side would grow more slowly or the lower side would grow more quickly. Alternatively, the growth rates of both sides might be affected. In this case there are three ways in which differential elongation of the two sides could be achieved (Firn & Digby 1980):

1. The elongation of both sides could be stimulated, with growth of the lower side to a greater degree.

2. Both sides could grow more slowly, with the retarding effect being greater on the upper side.

3. Gravistimulation could retard growth of the upper side while accelerating growth of the lower side.

Only by careful measurement is it possible to say precisely what effect gravistimulation has on the growth rates of the respective sides individually.

It is quite easy to make relevant measurements in the case of a relatively large and fast

Gravitropic Studies

growing organ like the dandelion. To measure a scape's growth rate before gravistimulation, place the scape in its glass tube and, at a recorded time, accurately mark with a waterproof felt tip pen a 10-cm section of the part outside the tube. Keep the scape vertical and remeasure the marked section after one, two or three hours. A scape taken at the inflorescence bud stage can be expected to extend about 0.5 cm for a 10-cm length in between two and three hours. Once the vertical growth rate has been established, place the tube in a horizontal position and make a note of the time when the scape starts to bend upwards (i.e,. when the lag time has elapsed). At precisely the same time, remeasure the distance between the two pen marks since some growth will take place during the lag period.

Let the scape bend until the tip just reaches a vertical position. Record the time and cut the scape at both pen marks. Some care should be taken to cut straight, at right angles to the sides of the scape. Split the scape longitudinally with a razor blade to separate the upper and lower sides. Lay each half flat, removing the bend with gentle pressure, and measure its length along the midline of the epidermal surface. It will immediately be apparent that the lower half is longer than the upper. Since the time required for bending is known, the growth rates of the upper and lower sides during tropism can be determined. We have always found that gravistimulation leads both to an increased growth rate of the lower side and a decreased growth rate of the upper side.

Presentation Time

When a scape is placed in a horizontal position and not subsequently disturbed, each part is continuously gravistimulated until it reaches a vertical position by bending. It has been found, however, that it is not necessary for an organ to be continuously gravistimulated to bring about a tropic response. In fact, the gravity stimulus need only be applied for a short period, known as the presentation time, for a tropic response to

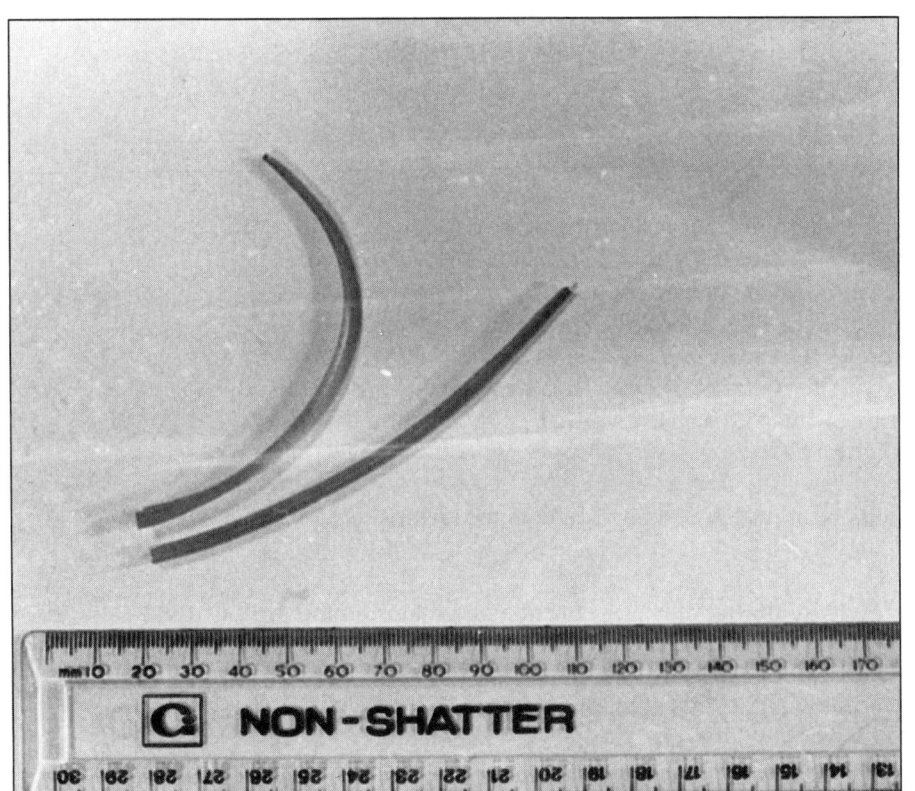

Figure 4. The 10-cm section of the scape is cut through at both ends and split down the middle to separate the upper and lower halves. When separated, the upper half of the scape bends even more due to release of tension.

be induced. It is quite easy to obtain an estimate of the presentation time (i.e. the minimum amount of time of exposure to gravity that will subsequently cause a bending response) for the dandelion scape.

Mount a series of scapes in glass tubes in a vertical position. Place each tube in a horizontal position for a different period of time ranging from 1-20 minutes, and immediately return each tube to the vertical position when the time is up. If the period of gravistimulation has been as long as or longer than the presentation time for the tissue, the scape will bend sideways toward the side that was uppermost when the scape was laid flat. Bear in mind, however, that there will still be a lag period and that no movement can be expected until this time has elapsed. Also, the scape, when returned to a vertical position, will be subjected to the gravity stimulus acting longitudinally. It will not continue to bend sideways for long but will tend to realign itself to its original, vertical orientation. Any sideways movement of the tip of the scape, however, should be counted as an indication that the scape was placed in a horizontal position at least for the presentation time for the material.

Gravity Detection

Plant organs are believed to perceive gravity by means of the sedimentation of amyloplasts within specialized cells called statocytes. The subject of graviperception has been well reviewed in this journal (Moore 1984). Dandelion scapes have particularly large amyloplasts and statocytes; for this reason, they are a good means for observing the gravity-detecting mechanism. The statocytes are located in the endodermis, particularly in regions adjacent to vascular bundles, and can easily be seen in transverse sections of the scape that have been stained with iodine. The amyloplasts, when stained in this way, show up well as black specks within endodermal cells, even in fairly crude hand-cut sections.

Figure 5. The upper and lower sides of the scape are laid flat and measured along the midline. The lower side has grown more than the upper side.

Gravitropic Studies

What makes the use of this material particularly exciting is that it is possible to observe unstained amyloplasts in living statocytes, and by appropriate manipulation, to watch them sedimenting under the influence of gravity.

To do this, take a dandelion scape that has been kept vertical at all times and use a new razor blade to cut median longitudinal sections by hand. Within these sections, provided they have passed close to a vascular bundle, will be statocytes with amyloplasts that have all settled on the basal wall. It is important to remember which is the basal end of each section so that when they are mounted for viewing with the microscope they are right side up. Mount the sections in water on a microscope slide with a cover slip so that their longitudinal axes are parallel to the sides of the slide. Until the slides are viewed through the microscope they should be kept upright with the basal ends of the sections pointing downwards. In this way the cells will be maintained in the same position as they were in the scape before the sections were cut.

Carefully turn the microscope on its side and fix the slide to the stage without altering the orientation of the sections. Find the statocytes with their amyloplasts under the 10× objective and then use the 40× objective to examine them in more detail. If students have

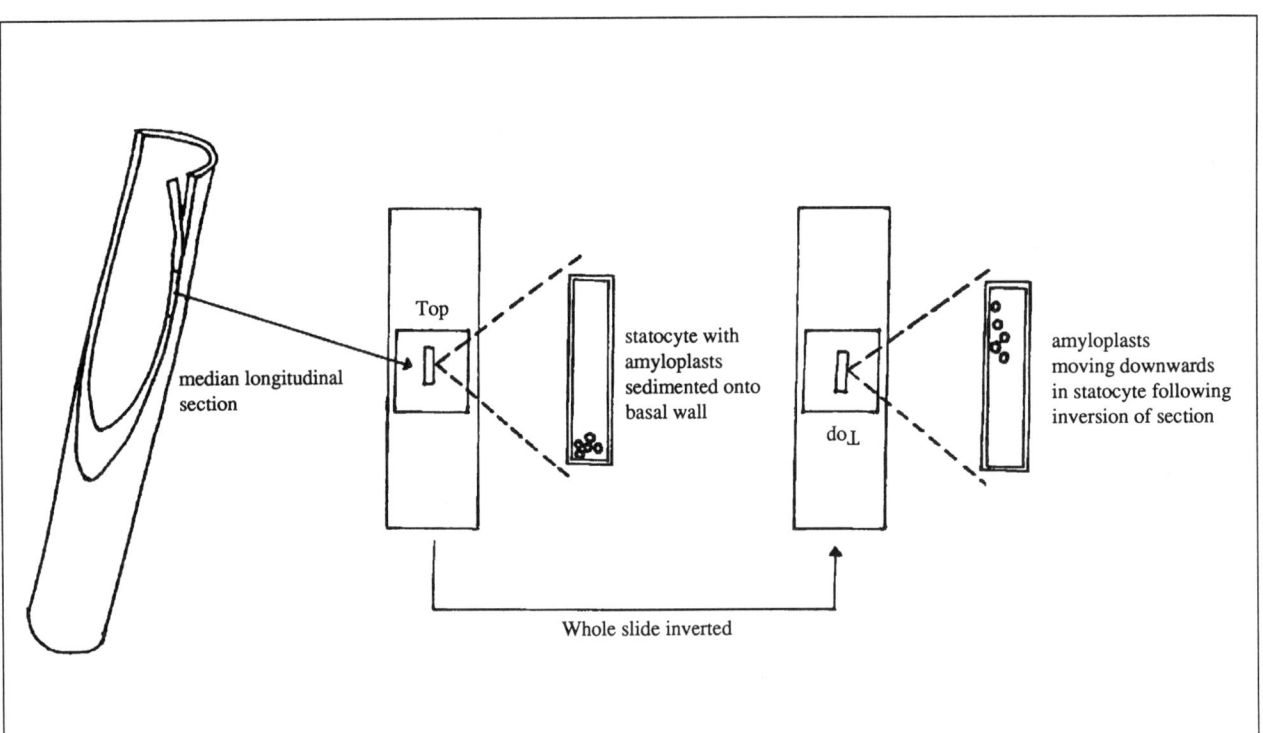

Figure 6. Procedure for watching amyloplasts sedimenting the living statocytes in a median longitudinal section of the dandelion scape.

difficulty finding the amyloplasts in unstained sections, using iodine-stained sections should help them know what to look for. It is possible that a section has not passed through any statocytes, but the xylem elements of the vascular bundles are a good clue to their presence.

Once a statocyte has been found, the amyloplasts identified and their location at the base of the cell confirmed, completely invert the slide (i.e., turn it through 180°) so that the basal wall of the statocyte, where the amyloplasts are clustered, is now uppermost. Do this either by revolving the microscope stage or, if this is not possible, by turning the whole microscope onto its other side.

The amyloplasts should now gradually sediment through the statocyte and their progress can be observed and recorded with a series of sketches to show their position in the cell. The use of a calibrated eyepiece graticule makes it possible to estimate their speed of movement.

As students watch amyloplasts moving in response to gravity, their attention is bound to be intensified if they are made aware that they are involved first-hand with one of the big mysteries of plant physiology. Much is known of the mechanism of gravitropism, but the link between graviperception by means of sedimenting plastids and resultant growth responses that bring about curvature is still unresolved.

References

Firn, R.D. & Digby, J. (1980). The establishment of tropic curvatures in plants. *Annual Review of Plant Physiology, 31*, 131-148.

Moore, R. (1984). How roots perceive and respond to gravity. *The American Biology Teacher, 46*(5), 257-265.

Oxlade, E.L. & Clifford, P.E. (1981). Experiments in geotropism. *Journal of Biological Education, 15*(2), 137-142.

Wilkins, M.B. (1984). Gravitropism. In M.B. Wilkins (Ed.), *Advanced plant physiology* (pp. 163-185). Marshfield, MA: Pitman.

Thin Layer Chromatography (TLC) of Chlorophyll Pigments

Jerry Foote, University of Wisconsin—Eau Claire, Eau Claire, Wisconsin

Chromatography is a rapid method of separating and identifying the components of a mixture without destroying it (Hechtlinger 1971). Several types of chromatography are now used with carbohydrate, amino acid, nucleic acid and lipid substances.

The use of various chromatographic techniques is increasing in today's high school classrooms. Klein (1981) proposes the use of paper chromatography as a basis for hypothesis formation, and Heck and Haworth (1974) discuss the separation of inks using this method. The separation of plant pigments by paper chromatography is described in most high school biology textbooks, and laboratory procedures for the process are outlined in most high school biology manuals. Klein (1979) suggests methods to separate plant pigments including a paper chromatographic technique using carbon tetrachloride as the solvent. Column chromatography of plant pigments is described by Dobbins (1972), and Heim (1970) describes a simple, inexpensive device for column chromatography in high school. Thin layer chromatographic techniques are used for food dye identification (McKone & Bell 1979; McKone & Nelson 1976), and separation of the components of Darvon capsules (Chasar & Toth 1974).

Several types of chromatography are complex processes requiring a number of organic solvents, expensive equipment, and long periods of time. These techniques are, thus, not usually done in the high school classroom. Paper chromatography of plant pigments is an easy, fairly fast process which can be accomplished within most high school science class periods and is undoubtedly the most widely used of the separation techniques. Thin layer chromatography (TLC) of plant pigments can also be easily done in the high school classroom, and the separation takes place much faster than with paper chromatography.

TLC is superior to paper chromatography for the separation of amino acids and nucleotides. The greatest advantages of TLC are excellent sharpness of separation, high sensitivity and speed. Separations that may require many hours on paper can be accomplished in a few minutes with suitable thin layers (Randerath 1968). Because TLC is an important biological technique, it should be demonstrated to high school biology students.

Disadvantages of doing TLC in high schools include: 1) the use of complex mixtures of toxic and inflammable solvents, and 2) the time and expense necessary to make the TLC plates. The solvent suggested for TLC by McKone and Bell (1979) is a mixture of four parts isoanyl alcohol, four parts 95% ethanol, one part concentrated aqueous ammonia, and two parts water. McKone and Nelson (1976) used a mixture of 50 parts n-butanol, 25 parts ethanol, 25 parts water, and 10 parts concentrated ammonia. Hechtlinger (1971) describes a procedure for TLC of chlorophyll pigments using a mixture of two parts isooctane, one part acetone, and one part diethyl ether for the solvent. Chaser and Toth (1974) describe the separation of the contents of a Darvon capsule using chloroform and concentrated ammonia, then extracting the ammonia portion with more chloroform, and finally chromatographing each of the chloroform solutions. For many high school biology teachers, these complexities, and the time necessary to complete them, will rule out their use.

TLC of Chlorophyll Pigments

With the following procedure, these disadvantages can be overcome and TLC using plant pigments can be done as easily as paper chromatography. If the two techniques are done together, they can both be accomplished in a single class period. The TLC procedure can be completed during the time the pigments are separating on the paper. Students can then learn the technique for both paper and thin layer chromatography. This procedure uses readily available materials, is simple to accomplish, has the *student do* each step of the process, and uses the same solvent chemicals as in paper chromatography.

Materials (Figure 1)

- Acetone
- Baby food jars
- Dried leaves
- Microscope slides
- Pasteur pipets
- Petroleum ether
- Silica gel H
- Stirring rod

Preparation of TLC Plates

The preparation of the TLC plates (microscope slides) is similar to that of Chasar and Toth (1974) and Hechtlinger (1971). To make the silica gel slurry, however, chloroform has been replaced by acetone.

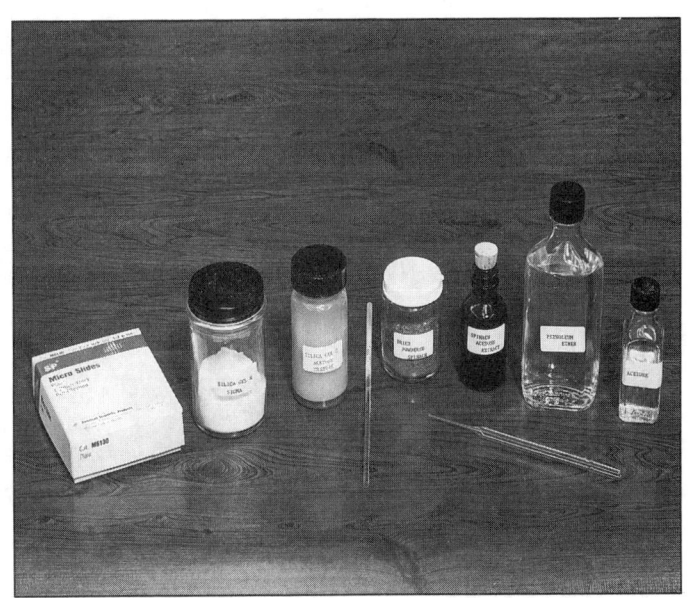

Figure 1. Materials necessary for TLC of chlorophyll pigments.

1. Mix 20 g of silica gel H with 70 ml of acetone in a glass bottle large enough to hold two microscope slides (a baby food jar works well). This amount of slurry will coat at least 20 slides. If the slurry becomes thick due to evaporation of the solvent, add a few more ml of acetone and stir again.

2. Stir the mixture well with a glass stirring rod.

3. Place two glass slides back to back with one slightly higher than the other. Dip the slides into the slurry and remove slowly touching the bottom of the slides to the edge of the container to drain off excess slurry (Figure 2). Carefully separate the slides and set them aside to dry. Drying will take only 10 or 20 seconds.

Biology Labs That Work: The Best of How-To-Do-Its

TLC of Chlorophyll Pigments

Preparation of Leaf Extract

4. Pulverize 2.5 g of dried leaves with a mortar and pestle. Fresh or frozen spinach leaves work well if dried for 48 hours in an oven or incubator at 37° C. Alternatively, commercially available dried parsley may be used. If an oven or incubator is not available for drying the leaves, fresh leaves may be macerated with a mortar and pestle. Maceration is more difficult with fresh leaves but the pigment separation, #9 and #10, occurs in a similar manner.

5. Mix the 2.5 g of pulverized leaves with 19 ml of acetone. Filter this mixture through filter paper into a dark colored, easily stoppered bottle. If fresh leaves are used, mix 10 ml of acetone with 2.5 g of macerated leaves. **Caution: Acetone is a toxic and extremely inflammable substance.**

Placing Extract on Slide

6. Place a micropipet or pasteur pipet into the leaf extract and allow the solution to move into the pipet by capillary action.

7. Place the pipet on the silica gel about 1 mm from the bottom of the slide allowing the solution to run onto the gel (Figure 3). The spot of solution will quickly dry. A second spot should be added and allowed to dry. Continue this procedure until a dark green spot is obtained (about 10 or 12 drops).

Figure 2. TLC ready for spotting.

For demonstration purposes and purely qualitative separations, large micropipets such as pasteur pipets work quite well. For more refined separations, true micropipets can be made by drawing out glass tubing into extremely thin capillaries.

Developing the Slide

8. Place 1.5-2.0 ml of chromatographic solution, a mixture of 4.5 parts of petroleum ether and one part acetone, into the chromatojar. The chromatojar should be slightly larger than the microscope slide and should have a tight cover (a baby food jar works well). Again, caution: This solution is toxic and very inflammable.

9. Place the slide in the chromatojar containing the solution and quickly cover. Separation of the pigments will occur in three or four minutes as the solvent moves up the slide (Figure 4).

10. When the solvent front nears the top of the silica gel, remove the slide from the chromatojar and allow to dry. Drying will take about 10 seconds. The chromatojar and solution may be reused many times.

11. Observe the pigments that have separated on the slide. As with paper chromatography, carotenes will be at the top of the slide; lutein, a xanthophyll, appears next as a dark line; the bluegreen color of chlorophyll a will be seen next; and finally near the bottom of the slide a yellow green mixture of chlorophyll b and xanthophylls is seen (Figure 5).

Discussion

This simplified procedure should work well in the high school classroom since the chemicals used are the same as those used in paper chromatography of plant pigments. Also, the procedure can be completed in the same laboratory period with paper chromatography since it can be done while the paper chromatography is developing; no need to find an extra class period for this procedure.

Changing the concentrations of the petroleum ether/acetone solvent mixture may give a better separation of some pigments. Varying the solution between 8:1 and 3:1 petroleum ether/acetone has produced good results but may separate some pigments well and others not at all.

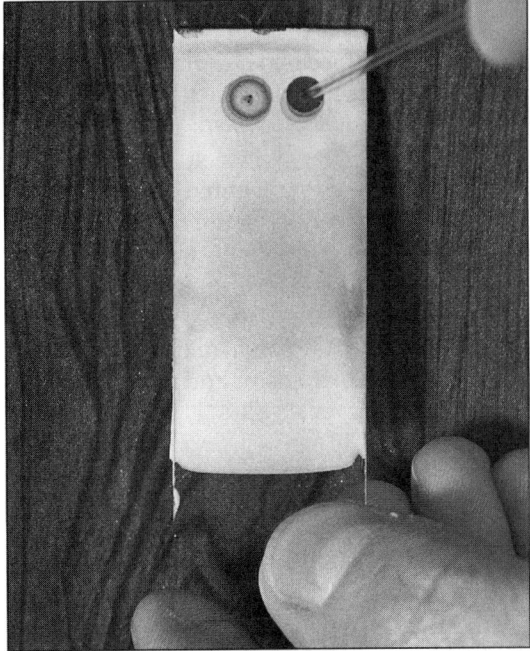

Figure 3. TLC plate being spotted.

Figure 4. Separation of pigments taking place in chromatojar.

TLC of Chlorophyll Pigments

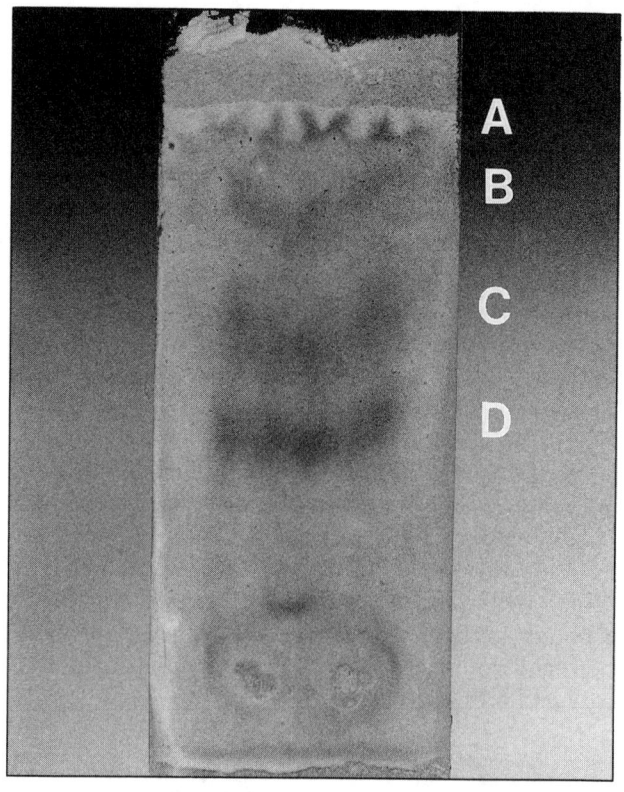

Figure 5. Pigments separated on TLC plate. A: Carotenes--Yellow color; B: Lutein--Gray color; C: Chlorophyll a--Bluegreen color; D: Mixture of chlorophyll b and Xanthophylls--Yellow-green color.

Some of the steps of the procedure can be deleted if the teacher so desires. This might save a bit of time but the student would not then actually see and do each part of the TLC procedure.

Preparation of the TLC plates (Steps 1, 2 and 3) can be circumvented by purchasing prepared TLC sheets. Using prepared TLC sheets might produce better results and save some time but would be more expensive if the process were done with very many students each year.

Entire kits may also be purchased. The use of a kit would save some mixing of solutions and preparing of the TLC plates but again would be more expensive and would prevent the students from actually taking part in each step of the TLC process.

References

Chasar, D.W. & Toth, G.B. (1974). Thin-layer chromatography of Darvon Compound-65. *Journal of Chemical Education, 51*(7): 487.

Dobbins, M.F. (1972). Column chromatography of plant and leaf pigments. *The American Biology Teacher, 34*(3): 160.

Hechtlinger, A. (1971). *Handbook of modern experiments for high school biology.* West Nyack, NY: Parker Publishing Co., Inc.

Heck, L.J. & Haworth, D.T. (1974). Disc paper chromatography of inks. *School Science and Mathematics, 74*(651): 3.

Heim, W.G. (1970). A simple inexpensive device for column chromatography. *The American Biology Teacher, 32*(7): 430.

Klein, R.M. (1979). Simple separation of plant pigments. *The American Biology Teacher, 41*(4): 240.

Klein, S.D. (1981). Why leaves turn color: A laboratory model for hypothesis formation. *The American Biology Teacher, 43*(3): 163.

McKone, H.T. & Bell, M. (1979). Identifying food dyes by chromatography. *Science Teacher, 46*(8): 37.

McKone, H.T. & Nelson, G.J. (1976). Separation and identification of some FD&C dyes by TLC. *Journal of Chemical Education, 53*(11): 722.

Randerath, K. (1968). *Thin-layer chromatography.* New York: Academic Press.

Rapid Germination of Pollen *In Vitro*

David J. Schimpf, University of Minnesota, Duluth, Minnesota

The growth of the pollen tube is the penultimate developmental event in the male gametophyte of angiosperms and gymnosperms, followed only by the discharge of sperm into the ovule. The tube grows by vacuolar enlargement and the extension of a very thin cell wall; usually the elongating vacuole is periodically truncated by transverse plugs of callose (Esau 1965). Although tube growth can take place *in vitro*, Sanders and Lord (1989) presented evidence that, by an unknown mechanism, cells within the style of the flower also bring about movement of the tip of the growing pollen tube of angiosperms.

Angiosperm species shed either binucleate or trinucleate pollen. The generative nucleus of a binucleate pollen grain undergoes mitosis after germination begins, yielding two sperm nuclei. The two sperm nuclei are already present in a trinucleate pollen grain at the time it is shed. Binucleate pollen is usually easy, but slow, to germinate *in vitro*. Trinucleate pollen germinates rapidly *in vivo*, but is generally difficult to germinate *in vitro*. It also loses viability much faster than does binucleate pollen. A few binucleate species are intermediate in character, germinating both rapidly and easily *in vitro*. Hoekstra (1979) reviews physiological differences among these types. Germination of the rapid binucleate types is suitable for observation during a single laboratory period.

The use of rapidly germinating pollen facilitates direct observation of an interesting aspect of the life cycle; it also lends itself well to experiments on the effects of environmental conditions on percentage of germination and rates of growth. In addition to being responsive to temperature, pollen tube growth is influenced by chemical substances in the germination medium (see e.g., Bilderback 1981). Wolter and Martens (1987) review the effects of air pollutants on pollen. The sensitivity of pollen to adverse environmental conditions has been used to predict the sensitivity of the sporophyte to the same adverse factors (Mulcahy & Mulcahy 1983).

Procedure

The medium that works best for germinating pollen is somewhat species dependent. One that many others have used has also worked well for me with the plants listed in Table 1: 10 g sucrose, 0.01 g H_3BO_3 and 0.03 g $CaNO_3$ plus distilled water to make 100 ml of solution. Germination usually occurs with only sucrose in the medium, but the percentage germination and rate of growth are often then reduced.

Pollen germinates poorly in a standard wet mount under a coverslip, which is probably the most frequent cause of disappointment with this exercise. Good germination occurred under a coverslip, though, when I used medium that was first saturated with commercial O_2 gas (>99.9% pure). This result suggests that the standard wet mount overly restricts diffusion of O_2 from the air to the pollen. Instead, use a drop of medium on a slide without a coverslip, a concavity slide with a coverslip or a hanging drop technique. If using a regular microscope slide and no coverslip, retard evaporation between observations by covering the slide with a petri dish lid that has moist paper stuck to its underside. If the preparation is on a concavity slide, leaving some air space under the coverslip sometimes enhances germination. Inexpensive chambers for hanging drops may be made by cementing coverslips to rubber cone washers, available where plumbing supplies are sold. Pollen can also be germinated on agar-based media coated onto slides. With some plants, pollen may be obtained by dabbing the anthers against the dry glass; with others the pollen is too adhesive or the anthers insufficiently dehisced for this, so the anther should be teased apart in the drop of medium on the slide. Grains on the outside of a large anther can be transferred with a needle, toothpick, forceps or small artist's brush to either dry glass or a drop of medium. Some kinds of pollen will remain viable for a few weeks if frozen in a dry, sealed vial. Germination is often enhanced with greater concentrations of pollen in the medium, probably because substances diffusing out of the grains promote germination. This will be evident through the microscope as greater germination percentages where there are clumps of pollen. However, extreme concentrations of pollen, such as those where grains occupy the entire field of view, seem to depress germination. Resource depletion or the buildup of diffusible inhibitors probably accounts for this.

How normal is *in vitro* development? Pollen tubes typically will not become nearly as long *in vitro* as they will *in vivo*. *In vitro* cultures may exhibit more than one pollen tube per grain. This could be an artifact of the technique, but multiple tubes are the *in vivo* norm for some taxa (McLean & Ivimey-Cook 1956).

I have had success with the taxa listed in Table 1. *Impatiens* and *Tradescantia* show especially rapid development, and *Impatiens* even germinates rather well in plain water. Pollen tubes are typically evident in 10-15 minutes and well developed in 30 - 45.

Acknowledgments

I thank Steve Flint for educating me about various aspects of pollen and Larry Hufford for comments on an earlier draft of the manuscript.

References

Bilderback, D.E. (1981). *Impatiens* pollen germination and tube growth as a bioassay for toxic substances. *Environmental Health Perspectives*, 37, 95-103.

Table 1. Some angiosperm taxa suitable for pollen germination in the classroom.

Catharanthus roseus (*Vinca rosea*) - Madagascar periwinkle
Eschscholzia - California poppy
    E. caespitosa, E. californica
Hemerocallis spp. - Day lily
Impatiens - Touch-me-not, jewellweed, cutivated impatiens
    I. capensis, I. wallerana
Malus spp. - Apple, crabapple
Papaver orientale - Oriental poppy
Prunus - Cherry (other groups within this genus not tested)
    P. cerasus, P. pensylvanica, P. virginiana
Rubus parviflorus -Thimbleberry
Tradescantia spp. - Spiderwort
Trifolium - Clover
    T. pratense, T. repens
Vinca minor - Blue or common periwinkle
Viola tricolor - Pansy

Rapid Germination *In Vitro*

Esau, K. (1965). *Plant anatomy* (2nd ed.). New York: John Wiley.

Hoekstra, F.A. (1979). Mitochondrial development and activity of binucleate and trinucleate pollen during germination *in vitro*. *Planta, 145,* 25-36.

McLean, R.C. & Ivimey-Cook, W.R. (1956). Textbook of theoretical botany (Vol. 2). London: Longmans, Green.

Mulcahy, D.L. & Mulcahy, G.B. (1983). Pollen selection: An overview. In D.L. Mulcahy & E. Ottaviano (Eds.), *Pollen: Biology and implications for plant breeding.* (pp. xv-xvii). New York: Elsevier Biomedical.

Sanders, L.C. & Lord, E.M. (1989). Directed movement of latex particles in the gynoecia of three species of flowering plants. *Science, 243,* 1606-1608.

Wolter, J.H.B. & Martens, M.J.M. (1987). Effects of air pollutants on pollen. *Botanical Review, 53,* 372-414.

Some Plant Hormone Investigations That Work

Donald S. Emmeluth, Fulton-Montgomery Community College, Johnstown, New York
Donald E. Brott, Fulton-Montgomery Community College, Johnstown, New York

Wouldn't it be nice to find some inexpensive laboratory investigations that are easy to set up and maintain, applicable at several levels of instruction and useful in several biology courses? If these investigations used living organisms showing constant directed change, illustrated both practical applications and creative thinking and were conducive to statistical analysis, then most teachers would probably be interested.

This article describes some plant hormone investigations that possess this type of potential. These investigations are usually performed simultaneously with discussion of growth regulating substances in the lecture portion of my course in plant biology. They could be set up prior to the lecture explanation and used as discovery exercises, or they could be performed in an introductory level course to illustrate the control and experimental conditions. What follows is the description of the investigations in the form given to students. They receive all of the information up to the section on *Additional Information*.

Hormonal Regulation of Plant Development

Introduction

Figure 1. Effect of Gibberelic Acid Treatment. Bean plants on left show the effect of gibberelic acid treatment after three weeks. Plants on the right are the untreated control.

The final form of a plant is a result of complex interaction involving the plant's genetic information and the modifying effects of its environment. Plant growth regulators are organic molecules that play a major role in modifying plant growth and development. A subgroup of plant growth regulators are the plant or *phyto*hormones. Like hormones in animal systems, they are produced in minute amounts and transported (translocated) to target sites where they cause some response. As cells differentiate and tissues and organs develop specific form and function, hormones provide a means of precise coordination between these individual parts during each stage of the organism's growth.

The control of differentiation and growth resides, ultimately, in the DNA of the nucleus. Nuclear information determines the production of hormones and other regulatory

Plant Hormone Investigations

chemicals and, in turn, may be subject to their modifying effects.

There are five widely accepted groups of plant hormones: auxins, gibberellins, cytokinins, ethylene, and abscisic acid. Research in this area is complicated because these phytohormones interact with each other in a number of different ways dependent on concentration, target site, species involved, developmental stage and environmental factors—to name but a few. Unlike animal or bacterial studies, there are usually no type species which may serve as general models for the effects of the phytohormones. Additionally, there are many other phytohormones restricted to specific plant groups and artificial plant regulators which may also fall into this classification.

The following investigations provide some insight into the effects of certain environmental and chemical factors on the patterns of plant growth and differentiation.

1. *Effect of Gibberellic Acid on Plant Growth.*

Select 20 plants (either pea or bean) that have been growing in soil for approximately two weeks. Divide the plants into two equal groups. Label one group "Control" and the other "Gibberellic Acid" or "GA3." Label each individual plant within each group with a number and the date or use a floor plan method. Measure and record the height of each plant (in mm) from the cotyledons to the tip of the shoot apex.

Apply one drop of gibberellic acid to the shoot apex of each plant in the tray labeled "Gibberellic Acid" and apply one drop of distilled water to the shoot apex of each plant in the Control group. **Be sure to use separate medicine droppers for each solution. Be sure to water the root systems of the plants with equal amounts of tap water.** Follow this procedure weekly, and record the height of each plant in each group and the general appearance of all the plants. When the experiment is concluded, plot the data (height vs. time), and determine whether growth response obtained with gibberellic acid is significantly different from that of the controls.

Figure 2. Effect of B-Nine Treatment. Marigold plants on the left have been treated with B-Nine for three weeks and illustrate reduced height due to disruption of IAA production. Plants on the right are the untreated control.

2. *Inhibitory Effects of Synthetic Plant Regulators.*

Succinic acid 2,2 dimethyl hydrazide or N-dimethylamino-succinamic acid (commonly known as Alar or B-Nine) is a synthetic plant regulator with effects directly opposed to the gibberellins and IAA. Choose five healthy tomato or marigold plants that have been growing in soil for three weeks to serve as your experimental group and five others to

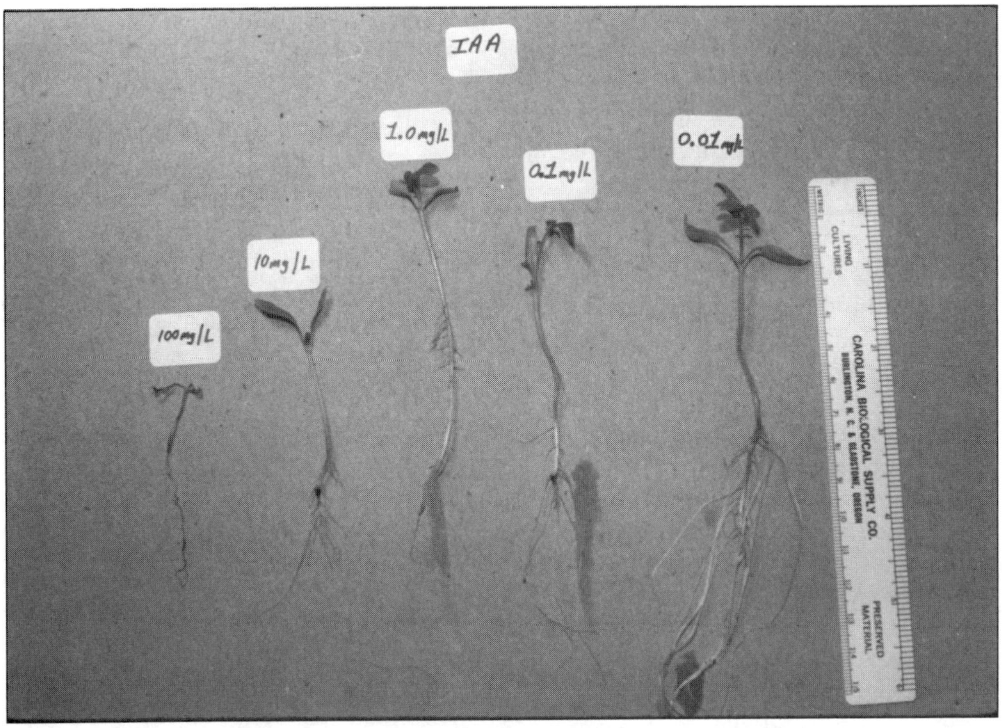

Plant Hormone Investigations

Figure 3. Inhibitory Effects of IAA on Root Growth. Tomato seedlings exposed to varying concentrations of IAA show reduced root growth. Plant exposed to 100 mg/liter concentration has died.

serve as controls. Label, measure and record the data for each plant in each group. Spray each plant in the control group with distilled water once a week for three weeks. Spray each plant in the experimental group with B-Nine once a week for three weeks. **Use separate atomizers for each group. Be sure to water the root systems of the plants with equal amounts of tap water.** During this time, note and record changes in height and general appearance of all plants in each group. Plot individual and collective differences and determine if height differences are statistically significant.

A useful follow-up is to plant the experimental and control groups in a garden or greenhouse and determine how long it takes for the experimental group to catch up to the control group.

3. *Effect of Varying IAA Concentrations on Root Growth.*

Choose six healthy tomato plants that have been growing in soil for approximately three weeks. Carefully remove each plant from the soil and gently wash the root system with tap water. Measure each plant and record its general appearance. Place each plant in a separate tube by threading the wet root system through the hole in the cork stopper. Fill each tube with the appropriate concentration of IAA so that the entire root system

Plant Hormone Investigations

is covered. Place aluminium foil or a similar material around the bottom of the tube to prevent the entrance of light. Place a sixth plant in a tube of tap water as a control. Observe root growth for the next three weeks.

Your instructor will show you how to compute the differing concentrations to be used (0.01, 0.10, 1.0, 10.0, 100 mg/liter).

Set up another set of tubes, using the same concentrations, with ACETIC ACID. Compare the results to those of IAA and control. Explain the results.

Additional Information

There are some general precautions to be considered when using any plant growth regulator and some specific considerations for each one used. Avoid contact with the skin, eyes and clothing. Hands should be washed thoroughly after using any of these chemicals. Keep all formulations refrigerated when not in use. If solutions are to be used for several weeks, make new ones about every two weeks. Be sure that all glassware is thoroughly washed and rinsed after contact with these chemicals.

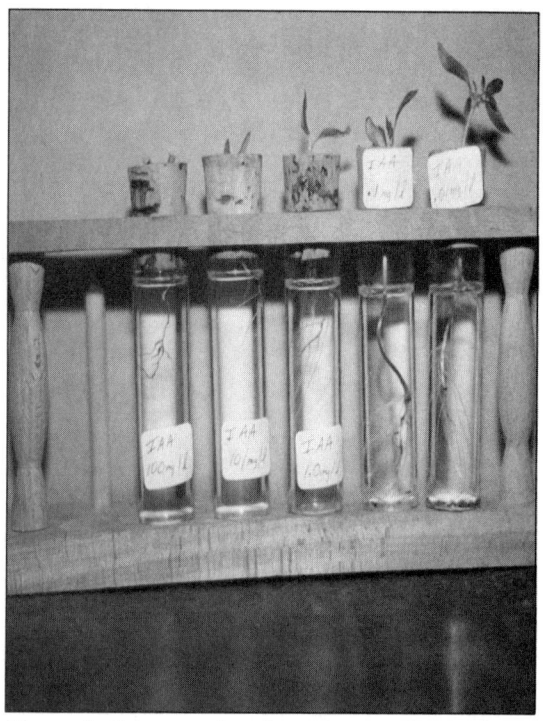

Figure 4. Setup used to demonstrate effects of differing IAA concentrations on root growth.

For Investigation I, the gibberellic acid solution is made by dissolving 100 mg of gibberellic acid powder in 2.5 ml of 95% ethyl alcohol. Dilute the resulting solution with distilled water to make 1 liter. This will give a solution of 100 mg/liter gibberellic acid. Almost any species of pea, bean or sunflower will work in this investigation. If you wish to show growth acceleration, "Little Marvel" dwarf peas are suggested. We have been successful with pole beans (see Figure 1). Since gibberellic acid may retard root growth, be careful not to spill any on the soil. The developing leaves of plants treated with gibberellic acid will start out as pale green and darken after the growth spurt is over.

B-Nine disrupts the synthesis of IAA and will tend to reduce internodal elongation. We have used several different species of tomato and marigold plants successfully (see Figure 2). When spraying the plants, cardboard shields should be used to prevent drifting of the solution. Wet the entire plant but try to prevent dripping or runoff from the plant. To prepare the B-Nine solution, dissolve one gram of B-Nine powder in 99 ml of distilled water. This provides a solution of 10,000 ppm, which is equivalent to a 1% solution. A wetting agent, such as Tween 20, should be added at a rate of one drop/50 ml of solution.

The purpose of Investigation III is to show the inhibitory effects of high IAA concentrations on root growth. While high concentrations of IAA will stimulate root formation, they tend to inhibit root growth in established roots. We have had good success using tomato seedlings (see Figure 3). Test tubes may be used to hold the plants. We use small vials which measure 83 mm by 22 mm. These vials require a #3 cork with a hole drilled by a #2 cork borer. If the plant is too small for the opening, wrap it with nonabsorbent cotton. Replace solutions as necessary. To prepare the IAA solution, dissolve 100 mg of IAA in 1.5 ml of absolute ethyl alcohol and add approximately 900 ml of distilled water. Cautiously warm the mixture on a hot plate to evaporate the alcohol and then dilute with

Plant Hormone Investigations

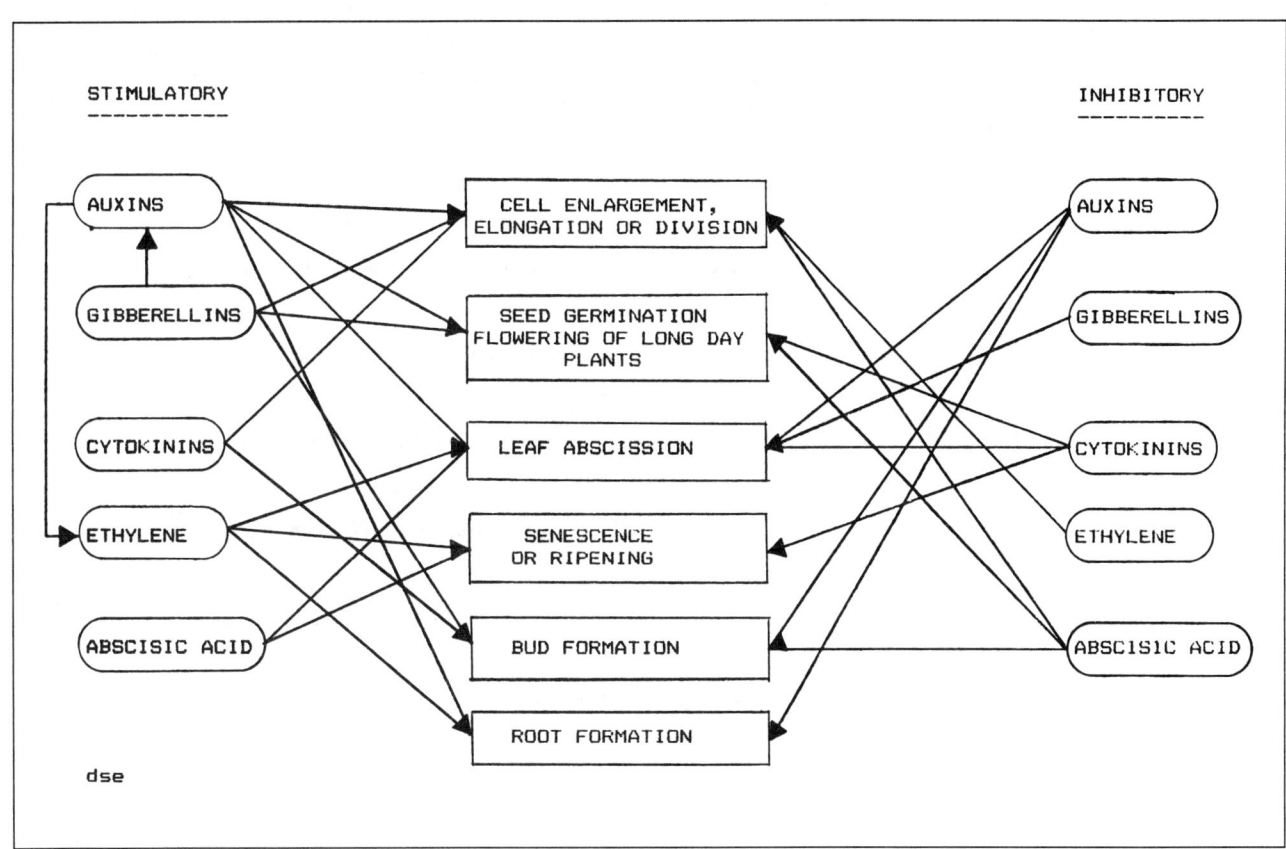

Figure 5. Generalized Reactions of the Major Plant Hormones. Virtually all hormonal reactions are affected by environmental stimuli such as light, temperature, water and carbon dioxide availability, pH, and presence of specific minerals or organic compounds. Reactions may differ from species to species, from tissue to tissue. Hormonal concentration often determines if the effect is stimulatory or inhibitory.

Biology Labs That Work: The Best of How-To-Do-Its

Plant Hormone Investigations

distilled water to make one liter. This provides a basic stock solution of 100 mg/liter. Each new dilution should be made from the preceding one. You'll see your best results near the end of the second week. If you plan to analyze your results statistically, be sure that you have a sufficiently large sample number. You will need to run several sets of each investigation to ensure statistically valid results.

Figure 5 shows some of the interactions among the five major plant hormone groups. It illustrates the concept of homeostasis and is given to students when we are discussing plant growth regulators in lecture.

Some Overlooked Objectives and Outcomes

It has been suggested (Mills 1981) that one major advantage of laboratory investigation is that it not only enables students to learn biology by doing, but also encourages creativity, objectivity, thoroughness and precision. I would suggest that other outcomes also occur while doing these investigations. Students learn the advantages and limitations of controlled experimentation. In addition, the instructor is challenged to keep abreast of information regarding plant growth regulators in order to answer the questions such investigations generate.

References

Abramoff, P. & Thomson, R.G. (1982). *Laboratory outlines in biology-III*. San Francisco, CA: W.H. Freeman and Company.

Evert, R.F. & Eichhorn, S.E. (1976). *Laboratory topics in botany* (2nd ed.) New York: Worth Publishers, Inc.

Flagg, R.O. & Stone, H.J. (1983). *Plant growth regulators*. Burlington, NC: Carolina Biological Supply Company.

Hereda, D. & Fieldhouse, D.J. (1970). Control of internal growth factors. *The Science Teacher, 37*(3), 83-85.

Lee, A.E. (1963). *Plant growth and development, A laboratory block*. Boston, MA: DC Heath and Company.

Mills, V.M. (1981). The investigative laboratory in introductory biology courses: A practical approach. *The American Biology Teacher, 43*(7), 364-367.

Animals

Artificial Urine for Laboratory Testing

Brian R. Shmaefsky, Kingwood College, Kingwood, Texas

The fear of contracting contagious microbial agents through body fluids has led to laboratory practices that limit exposure to blood, exudates, saliva and urine (Sharp & Smailes 1989). Due to the nature of the techniques and materials handled, clinical teaching laboratories require the utmost protection for students. Students are inexperienced at handling hazardous substances and are at risk of being accidentally contaminated with potentially fatal pathogens: hepatitis A and B, HIV, and sexually transmitted bacteria (Ballman 1989). Elimination of the clinical testing of body fluids from an anatomy and physiology curriculum is not recommended since students must be familiar with the psychomotor aspects of clinical testing before taking a job in the health sciences. Thus, body fluid substitutes must be found that will permit students to have the necessary clinical experience while preventing their accidental inoculation with a contagious disease.

Artificial blood and urine for use in standardizing clinical equipment is available from some specialty chemical companies, but these are generally too costly for the limited budgets of many high schools, undergraduate colleges and patient education programs. Some biological supply companies market bottles of crude artificial urine for eliciting normal and positive tests from simple urine assays. These products are clinically satisfactory for school use and are not very expensive. A major limitation is that the instructor does not have control over the test result outcomes. A simple-to-prepare and inexpensive artificial blood has been devised for blood-typing in the classroom (Sharp & Smailes 1989). This allows students to safely practice performing simple chemical experiments on blood without the danger of contamination. The urinalysis laboratory is equally important as blood-typing and could be safely conducted using instructor-prepared artificial urine. The artificial urine described in this article could be used to study various urine parameters using a urine hydrometer for specific gravity and urine test strips (dipstick) for the measure of glucose, ketones, pH and proteins. Procedures are included for test strip measures indicating the presence of abnormal urine parameters, erythrocytes and leukocytes, if the instructor wishes to use these tests.

Materials

The following reagents will be necessary for the preparation of normal human urine (Kark et al 1964):

- Albumin powder (egg or bovine)
- Creatinine
- Distilled water
- Potassium chloride
- Sodium chloride
- Sodium phosphate (monobasic)
- Urea

Procedure

Artificial Urine

Normal Human Urine

A class of 30 students, working in groups of two, would require a class total of at least 1 liter of artificial urine for specific gravity and dipstick testing. The following instructions are for the preparation of approximately 2 liters of normal urine; half can be stored or used for abnormal urine studies.

To 1.5 liters of distilled water add 36.4 g of urea and mix until all the crystals are dissolved. Then add 15.0 g of sodium chloride, 9.0 g of potassium chloride and 9.6 g of sodium phosphate; mix until the solution is clear. Check the pH with indicator paper or a pH meter to ensure the pH is within the 5 - 7 pH range for normal urine; if the solution is out of this pH range, the pH may be lowered with 1N hydrochloric acid or raised with 1N sodium hydroxide.

Next, place a urine hydrometer into the solution and dilute with water until the solution is within the specific gravity range of 1.015 to 1.025. This solution will serve as the storage stock solution of "normal urine solution" and may be kept refrigerated for several weeks or frozen in plastic containers for months. Before use, the stock solution should be warmed to room temperature. Then, to ensure a similarity to human urine, 4.0 g of creatinine and 100 mg of albumin may be slowly mixed into the 2 liters of the so-called normal urine solution.

Abnormal Human Urine

The artificial human urine may be modified to mimic several diseased or periodic conditions that are detectable in the urine. Abnormal urine is normally not available from student samples, thus students rarely experience the test results associated with disease. Also, the artificial abnormal urine is an excellent medium to test student skills and observations with the clinical testing of urine. The following conditions may be exhibited by using the "normal urine solution" and additional inexpensive reagents.

1. **Glycosuria**: High levels of glucose due to diabetes mellitus, pregnancy, excessive stress, renal tubular damage and brain damage. Add a minimum of 600 mg of glucose (dextrose) to each liter of "normal urine solution" to obtain a minimally detectable level of glycosuria. A moderate to high level of glycosuria can be achieved by adding 2.5 to 5.0 g of glucose to each liter of the solution. Sucrose or other sugars will not substitute for glucose; only glucose yields positive results with most urine test strips. Vitamin C (ascorbic acid) contamination of the urine, at values of 400 mg/l or greater, though, does yield false positive glucose results. This may add an interesting twist to the study of the accuracy and limitations of urinalysis.

2. **Proteinuria**: A high level of protein in the urine is an excellent indicator of glomerular damage. In the absence of glomerular damage, elevated urine protein

Artificial Urine

may result from excessive exercise, cold exposure and acute abdominal diseases. Protein levels in excess of 300 mg of albumin per liter of "normal urine solution" will give positive results. Severe renal damage may be exemplified by adding 1 g of albumin to each liter of the urine solution.

3. **Ketonuria**: Ketones of various types, which are normal liver metabolites, should not be found in detectable amounts in the urine. Elevated ketone levels are indicative of cold exposure, diabetes mellitus, dietary imbalances, and genetically or chemically acquired metabolic abnormalities. Ketonuria may be exhibited by adding a minimum of 100 mg of acetacetic acid or at least 1 ml of acetone to 1 liter of "normal urine solution."

4. **Urine pH imbalances**: Acidic urine can be obtained by adjusting the pH of the "normal urine solution" to a pH of 4.0 to 4.5 with 1N HCl. Consistent acidic urine is a sign of metabolic or respiratory acidosis, methanol poisoning, or metabolic disorders (for example phenylketonuria). Alkaline urine is obtained by adjusting the pH of the "normal urine solution" to a pH of 8 - 9 using 1N NaOH. Consistent alkaline urine is indicative of metabolic and respiratory alkalosis and urinary tract infections.

5. **Hyposthenuria**: Urine should have a specific gravity range of 1.015 - 1.025; some daily variation outside of this range is normal. Consistent production of dilute urine, with a specific gravity less than 1.015, is an indication of cardiovascular problems, diabetes insipidus, or renal tubule problems. The specific gravity of the "normal urine solution" may be lowered by adding distilled water to a volume of stock solution until the specific gravity approaches 1.005.

6. **Hemoglobinuria**: Hemoglobin in the urine results from excess levels of free hemoglobin in the blood due to excessive red blood cell lysis, renal damage, or normal menstrual flow. Bovine (cow) hemoglobin is an inexpensive powdered reagent available from many biological and chemical supply companies. Hemoglobinuria can be exhibited by adding 260 mg of bovine hemoglobin to 1 liter of "normal urine solution." Hematuria, or the presence of whole blood in urine (a good indication of glomerular damage), may be modeled using heparinized or defibrinated sheep blood normally used in microbiological and cell cultures. The urine test strips are normally sensitive to 1 ml of whole blood in 1 liter of urine solution.

7. **Leukocyte presence**: This is a difficult test and requires the use of small amounts of reagents and urine. The presence of leukocytes in urine indicates urinary tract damage or infection. The urine test strips can be faked into giving a positive leukocyte response by the addition of enzymes called esterases. Esterases are available through biological, chemical and histological supply companies. Esterase activity is measured in activity units; many companies sell the enzymes by the unit or by units per mass. A positive test for leukocytes may be achieved by adding 100 to 200 units of pork or rabbit liver esterase to 100 ml of the "normal urine solution." The leukocyte test must be performed immediately after the addition of the enzyme and may be performed on 10 ml samples in small test tubes.

A whole spectrum of urine abnormalities could be included in one sample by mixing the appropriate amounts of the "abnormal conditions reagents" into a common 1 liter volume of the "normal urine solution." For example, the urine from a patient with diabetes mellitus would have urine that tests positive for glycosuria and ketonuria, while a patient with glomerular damage would have urine that is positive for proteinuria, hemoglobinuria and hematuria.

The above method for the production of artificial urine is pragmatic for several reasons. It allows students to perform urinalysis without the fear of contamination by hazardous microorganisms. The procedure allows for the manipulation of the "urine" so students can encounter the diseased urine types not normally found among student urine samples. The materials for making the artificial urine are inexpensive and available from many biological and chemical supply companies. Lastly, the preparation is simple, and large quantities may be stored for several months. The artificial urine is only accurate for use with urine test strips or related reagents: It is not intended for use with electronic clinical analyzers.

Acknowledgments

I would like to thank Mary Jo Shmaefsky for her assistance with the formulation of the artificial urine chemistry. I would also like to thank Douglas Eder, who fostered my creativity in the teaching of biology.

References

Ballman, G. (1989). Handling infectious materials in the education setting. *American Clinical Laboratory, 8*(7), 10-11.

Kark, R.M., Lawrence, J.R., Pollack, V.E., Pirani, C.L., Muehrcke, R.C. & Silva, H. (1964). *A primer of urinalysis* (2nd ed.). New York: Hoeber Medical Division, Harper & Row, Publishers.

Sharp, R.H., Sr. & Smailes, D.L. (1989). A simulation of the blood type test. *The American Biology Teacher, 51*(4), 232-233.

The Sea Urchin Embryo: A Remarkable Classroom Tool

Steven B. Oppenheimer, The California State University, Northridge, California

The experimental material of choice for the investigation of many developmental mechanisms is the sea urchin embryo (Giudice 1986; Davidson et al 1982; Oppenheimer & Lefevre 1989). It is the organism of choice because specific qualities also make it ideal for classroom use. Unlike chick embryos, which are covered by shells and available in small numbers, and unlike frogs, whose males are often sacrificed to obtain sperm and whose females are hormonally induced to ovulate, sea urchin embryos are available by the billions and clearly display embryonic development. No shells are present to block viewing, and all experiments are done in the simplest of media—natural or artificial sea water. And, teachers can easily obtain sea urchins.

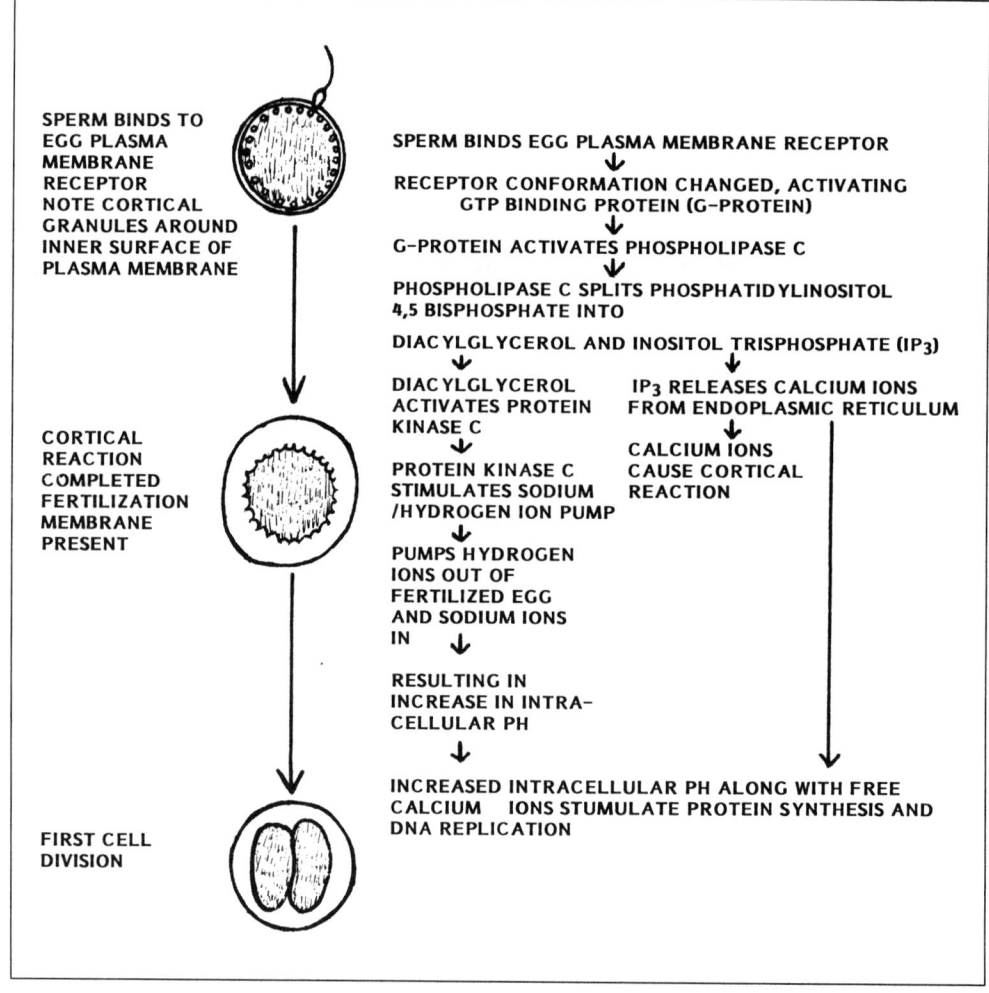

Figure 1. Model showing proposed causative sequence of some events occuring during egg activation in sea urchin fertilization.

Sea Urchin Embryo

Sea urchin embryos have been used for decades in the classroom and research laboratory and are the finest tools available for introducing students to the wonders of embryonic development and the world of research science. This article will illustrate how the sea urchin has been used to uncover key developmental mechanisms and how it can be used in the classroom to excite the students' curiosity and facilitate their introduction to well defined research experiences.

The Sea Urchin in Research

Many important discoveries in the areas of fertilization and early development resulted from experiments with sea urchin embryos. One such exciting discovery as the story of egg activation.

How does a tiny sea urchin sperm that is only 0.0002% of the egg surface trigger the multitude of changes that occur in the fertilized egg? The sea urchin has been most instrumental in answering this question. Within three seconds after the sperm binds to the sea urchin egg, a membrane potential change occurs (Whitaker & Steinhardt 1985). By 30 seconds, calcium ions begin to be released from the endoplasmic reticulum to a free state in the cytoplasm, followed by the cortical reaction, in which cortical granules that line the inner surface of the egg plasma membrane begin to fuse with the membrane, releasing their contents. This leads to the formation of the fertilization membrane that blocks the entry of additional sperm (Whitaker & Steinhardt 1985; Oppenheimer & Lefevre 1989) (Figure 1).

A variety of elegant experiments have led to the key finding that free calcium ions are directly responsible for the cortical reaction. Steinhardt, Epel, Chambers, Pressman and Rose used a substance called calcium ionophore A23187, which causes release of stored calcium in cells, duplicating some of the events occuring shortly after sperm binding. They found that many of the same events that occured after sperm binding also occured with the use of this chemical in the absence of sperm. This suggested that calcium ions must play a key role in egg activation (Steinhardt & Epel 1974; Whitaker & Steinhardt 1985).

This suggestion was strengthened by Victor Vacquier's experiments at the Scripps Institution of Oceanography. Sea urchin eggs were bound to glass slides and lysed, exposing the inner membrane surface to which the cortical granules are attached. The slides were exposed to a variety of salt solutions. Only calcium ions caused fusion of the cortical granules with the plasma membrane in much the same way as in the intact egg during the cortical reaction. Experiments such as these, using sea urchin material, have been instrumental in helping us to understand some of the events that occur during fertilization (Vacquier & Epel 1978).

Recent work with the sea urchin system has helped explain exactly how sperm cause egg activation. The binding of sperm to the egg cell membrane receptor appears to change the conformation of the receptor, which activates a GTP-binding protein (G-protein)

Sea Urchin Embryo

(Turner et al 1986). This protein then activates phospholipase C, which in turn splits phosphatidylinositol 4,5 biphosphate into diaclyglycerol and inositol trisphosphate (IP3). IP3 causes the endoplasmic reticulum to release calcium ions, which in turn cause the cortical reaction to occur. Diacylglycerol activates protein kinase C, which stimulates the sodium/hydrogen pump to pump hydrogen ions out of the egg and sodium ions in, resulting in an increased intracellular pH. This rise in pH, along with the free calcium ions, appears to be instrumental in activating protein synthesis and DNA replication (Berridge 1985; Swann & Whitaker 1986; Ciapa & Whitaker 1986; Whitaker & Irvine 1984; Busa et al 1985; Gilbert 1987). Figure 1 summarizes the proposed causative events in sea urchin egg activation.

The early embryo now cleaves and develops into the hollow ball stage (blastula). This is followed by the gastrula, a stage in which many dramatic changes occur. The well known embryologist Lewis Wolpert is widely believed to have said that it is not birth, marriage, or death, but gastrulation which is truly the most important time in your life. During gastrulation, the embryo begins to take shape. Without this process, many organisms would be round little balls that could never amount to anything.

The sea urchin embryo, because of its simplicity and transparency, has been helpful in understanding gastrulation mechanisms. Investigators have been able to observe how the cells behave during gastrulation. Small cells, called micromeres, lose adhesive affinity with other cells in the vegetal plate region of the blastula. They migrate into the central cavity—the blastocoel—and are called primary mesenchyme cells. These cells migrate along the extracellular matrix in the blastocoel by tenaciously adhering to fibronectin, a large glycoprotein secreted by blastula cells (Wessel et al 1984; Fink & McClay 1985) and appears to control their migration (Katoh & Hayashi 1985). A variety of experiments in which synthesis of certain sulfated glycoproteins was inhibited or assembly of microtubules prevented suggests that these components also play important roles in mesenchymal cell migration (Karp & Solursh 1974; Anstrom et al 1987; Gibbins et al 1969).

In some sea urchin species, projections, called filopodia, which extend from secondary mesenchyme cells at the advancing tip of the primitive gut (archenteron), stick to the inner surface of the blastocoel wall and contract, helping to complete the formation of the elongated archenteron (Trinkaus 1984).

These are a few of the advances in developmental biology discovered through experiments with sea urchin embryos. Sea urchins can also be used successfully in the classroom.

Classroom Experiments

Sea urchin kits containing all the materials and instructions for experiments for 600 or more students are available commercially at a cost of about $130 (which includes shipping by air express). When the kit arrives, the sea urchins can be used immediately or stored in a refrigerator as packed for up to a few days. After gametes are removed from the sea urchins by inoculating them with 0.5 M potassium chloride (included in the kit), the undiluted sperm and the eggs diluted in sea water can be used immediately or stored for up to a few days in the refrigerator. Long term maintenance of adult sea urchins is best done in refrigerated marine aquaria at 9-12° C.

Large groups of students can be introduced to sea urchin fertilization by placing a small drop of eggs on a slide. As the student views the eggs under the microscope, a drop of freshly diluted sperm (0.1 ml undiluted sperm added to 10 ml sea water) is added and fertilization can be clearly viewed. A discussion of the events that occur during fertilization, as presented earlier (Figure 1), can provide students with a feeling for what is going on right before their eyes.

Early development can also be beautifully observed by students using this system (Figure 2). Fertilize the diluted eggs (1 ml settled eggs in 100 ml of sea water) in a large beaker with freshly diluted sperm. Allow the eggs to settle out; pour off the sea water/sperm suspension and refill with fresh sea water (natural or artificial). Pour the diluted zygotes into plastic petri dishes until the dishes are half full and store them in a cool room (15-17° C is best). In a couple of hours the zygotes will undergo cleavage and in about a day the blastula stage embryos will hatch out of their fertilization membranes; gastrulation follows.

Student Research

With the introductory exercise behind them, students are generally so intrigued with this living system that they are eager to do more. This system provides an ideal opportunity to introduce them to personal research projects. For nearly two decades we have been using these sea urchin fertilization and development exercises as an introduction to student research in both pre-college and college programs. These programs have received widespread recognition through NSF grants, an NIH grant, Thomas Eckstrom Trust and Joseph Drown Foundation grants, NASA grants, and fellowships and awards from the Trustees of the California State University (CSU) system, American Cancer Society and California Science Teachers Association.

What is so great about using the basic sea urchin in class is that offshoots are easily accomplished in minutes. First I describe possible projects that students can easily do in the classroom using little more than sea urchin gametes and artificial sea water. For example, student projects can involve changing salt concentrations of sea water, changing specific ions or adding chemicals, changing the pH or temperature of the sea water, and observing the effects of these changes on fertilization or early development. You can use criteria such as counting percent of eggs with fertilization membranes or abnormalities observed during development compared with normal control conditions.

These experiments are so simple to do and the results so easy to obtain that generally 100% of the students in a class are able to successfully carry out one of these mini-

Sea Urchin Embryo

projects. We require that a brief experimental plan with background references first be submitted to the instructor who then makes suggestions and returns it to the students. Students make up their solutions during one class period and conduct the experiments during the next couple of days. Upon completion of the project, students write up their experiments according to standard format found in science journals and follow it up with an oral presentation of their work. This approach generally works best when students work in groups of three or four where each student has specific tasks to accomplish. The group setting improves self confidence and leads to a high degree of mutual assistance, as has been found in other learning situations (Lapp et al 1989; Johnson & Johnson 1975).

Our students are working on some very successful projects, as are teachers participating in our National Science Foundation-sponsored teacher enhancement program. A high school student who has been working with us for two years studied the effects of direct electric current on fertilization and early development in the sea urchin. He won a best paper finalist award at the Southern California Academy of Sciences annual meeting. Another student has been awarded an $18,000-a-year fellowship from NASA to work with us on a computer analysis of the parameters affecting sea urchin fertilization and early development. This will prepare the sea urchin system for study under zero gravity conditions in space.

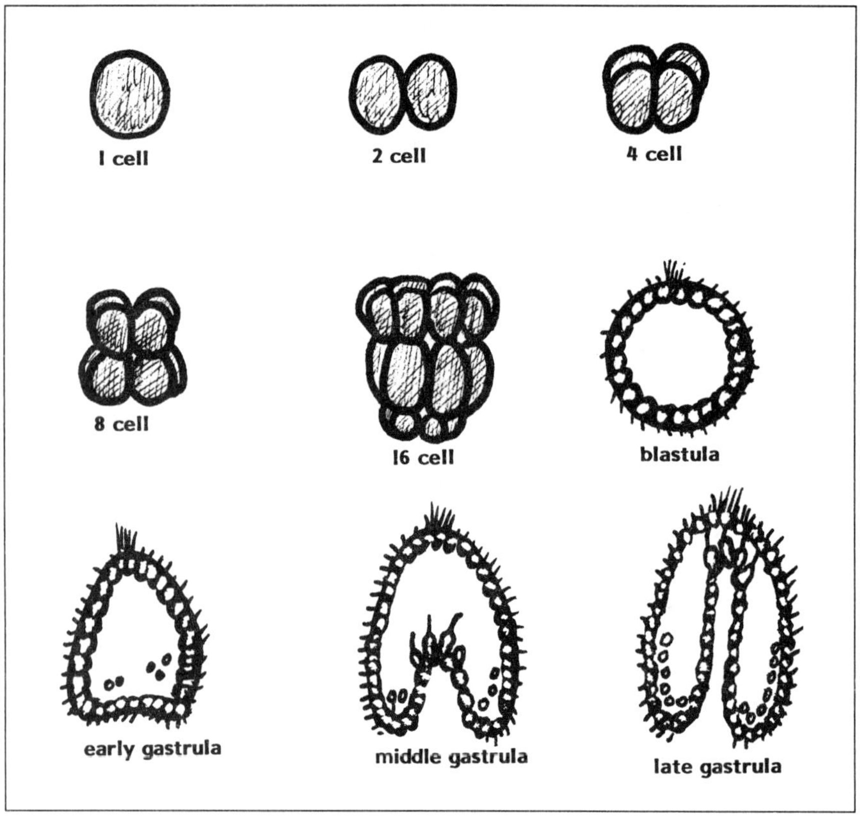

Scores of students in our laboratory are studying the molecular mechanisms involved in controlling cell adhesion in the sea urchin embryo (with support over the years from NSF, NIH, NASA, Thomas Eckstrom Trust, Joseph Drown Foundation, CSU Foundation, Northridge Student Projects Committee, and CSU Research and Grants Committee). In these experiments, students grow sea urchin embryos to the swimming blastula stage (about 23 hours for the sea urchin *Strongylocentrotus purpuratus*), then disaggregate them into viable single cells by incubating the embryos in calcium-magnesium-free sea water. When the cells are returned to normal sea water that

Figure 2. Cleavage and gastrulation in the sea urchin embryo.

contains calcium and magnesium, they reaggregate to form swimming embryo-like structures called embryoids, which can undergo further development. The students therefore learn that by simple experimental manipulations, embryos can be taken apart and put back together.

Our students use this intriguing concept to study the molecules required for cell adhesion, a property that is essential for normal embryonic development and which, when defective, plays a key role in the spread of cancer. They incubate the single sea urchin embryo cells with a variety of substances, some isolated from living sea urchin embryos, to test their effects on the ability of the cells to reaggregate. These simple student experiments have led to the discovery of specific proteins and sugars that appear to be involved in mediating adhesion in this system (Asao & Oppenheimer 1979; Oppenheimer & Meyer 1982a, 1982b).

Many students who were introduced to the sea urchin in our classes are now research scientists and have been co-authors of our research publications (28 of our research publications include 72 student co-author citations). The sea urchin embryo is so easy to work with and so likely to yield meaningful results that we believe it can't be beaten as a tool to introduce students to the excitement of experimental biology.

Full-page copies of the figures included in this article and other teaching aids for classroom use can be obtained free of charge by contacting the author.

References

Anstrom, J.A., Chin, J.E., Leaf, D.S., Park, A.L. & Raff, R.A. (1987). Localization and expression of msp 130, a primary mesenchyme lineage-specific cell surface protein of the sea urchin embryo. *Development, 101*, 255-265.

Asao, M.I. & Oppenheimer, S.B. (1979). Inhibition of cell aggregation by specific carbohydrates. *Experimental Cell Research, 120*, 101-110.

Berridge, M.J. (1985). The molecular basis of communication with the cell. *Scientific American, 253*, 142-152.

Busa, W.B., Ferguson, J.E., Joseph, S.K., Williamson, J.R. & Nuccitelli, R. (1985). Activation of frog (*Xenopus laevis*) eggs by inositol triphosphate. 1. Characterization of Ca^{2+} release from intracellular stores. *Journal of Cell Biology, 100*, 677-682.

Ciapa, B. & Whitaker, M. (1986). Two phases of inositol polyphosphate and diacylglycerol production at fertilization. *FEBS Letters, 195*, 347-351.

Davidson, E.H., Hough-Evans, B.R. & Britten, R.J. (1982). Molecular biology of the sea urchin embryo. *Science, 217*, 17-26.

Sea Urchin Embryo

Fink, R.D. & McClay, D.R. (1985). Three cell recognition changes accompany the ingression of sea urchin primary mesenchyme cells. *Developmental Biology, 107,* 66-74.

Gibbins, J.R., Tilney, L.G. & Porter, K.R. (1969). Microtubules in the formation and development of the primary mesenchyme of *Arbacia punctulata*. *Developmental Biology, 18,* 523-539.

Gilbert, S.F. (1987). *Developmental biology* (2nd ed.). Sunderland, MA: Sinauer.

Giudice, G. (1986). *The sea urchin embryo*. Boston, MA: Springer Verlag.

Johnson, D. & Johnson, R. (1975). *Learning together and alone: Cooperation, competition and individualization.* Englewood Cliffs, NJ: Prentice Hall.

Karp, G.C. & Solursh, M. (1974). Acid mucopolysaccharide metabolism, the cell surface, and primary mesenchyme cell activity in the sea urchin embryo. *Developmental Biology, 41,* 110-123.

Katoh, H. & Hayashi, M. (1985). Role of fibronectin in primary mesenchyme cell migration in the sea urchin. *Journal of Cell Biology, 101,* 1487-1491.

Lapp, D., Flood, J. & Thorpe, L. (1989). Cooperative problem solving, enhancing learning in the secondary science classroom. *The Science Teacher, 51,* 112-115.

Oppenheimer, S.B. & Lefevre, G. (1989). *Introduction to embryonic development* (3rd ed.). Boston, MA: Allyn and Bacon.

Oppenheimer, S.B. & Meyer, J.T. (1982a). Isolation of species-specific and stage-specific adhesion promoting component by disaggregation of intact sea urchin embryo cells. *Experimental Cell Research, 139,* 472-476.

Oppenheimer, S.B. & Meyer, J.T. (1982b). Carbohydrate specificity of sea urchin blastula adhesion component. *Experimental Cell Research, 139,* 451-455.

Steinhardt, R.A. & Epel, D. (1974). Activation of sea urchin eggs by a calcium ionophore. *Proceedings of the National Academy of Sciences, U.S.A., 71,* 1915-1919.

Swann, K. & Whitaker, M. (1986). The part played by inositol triphosphate and calcium in the propagation of the fertilization wave in sea urchin eggs. *Journal of Cell Biology, 103,* 2333-2342.

Trinkaus, J.P. (1984). *Cells into organs: The forces that shape the embryo* (2nd ed.). Englewood Cliffs, NJ: Prentice Hall.

Turner, P.R., Jaffe, L.A. & Felin, A. (1986). Regulation of cortical vesicle exocytosis in sea urchin eggs by inositol 1.4.5-triphosphate and GTP-binding protein. *Journal of Cell Biology, 102,* 70-76.

Vacquier, V.D. & Epel, D. (1978). Membrane fusion events during invertebrate fertilization. In G. Poste & G.L. Nicolson (Eds.), *Membrane fusion*. New York: Elsevier.

Wessel, G.M., Marchase, R.B. & McClay, D.R. (1984). Ontogeny of the basal lamina in the sea urchin embryo. *Developmental Biology, 103,* 235-245.

Whitaker, M. & Irvine, R.F. (1984). Inositol 1,4,5-triphosphate microinjection activates sea urchin eggs. *Nature, 312,* 636-639.

Whitaker, M.J. & Steinhardt, R. (1985). Ionic signalling in the sea urchin egg at fertilization. In C.B. Metz & A. Monroy (Eds.), *Biology of fertilization* (Vol. 3, pp. 167-221). Orlando, FL: Academic Press.

Artificial Urine Test To Simulate the Test for Pregnancy

A.J. Russo, Mt. St. Mary's College, Emmitsburg, Maryland

At approximately the sixth day of pregnancy in humans, the developing embryo, now in its blastocyst form, attaches itself to the endometrium of the mother's uterus. The blastocyst, now only about one hundredth of an inch in diameter, absorbs nutrients from the glands and blood vessels of the endometrium for its subsequent growth and development. At about the same time the trophoblast cells begin to secrete human chorionic gonadotropin (HCG). The primary role of HCG seems to be to maintain the activity of the corpus luteum, which needs to continue to secrete progesterone for the continued attachment of the fetus to the endometrial wall. HCG is excreted in the urine of pregnant women from about the eighth day of pregnancy, reaching its peak of excretion at about the eighth week of pregnancy. The detection of HCG in the urine of pregnant women is the basis for all pregnancy tests, including the "over-the-counter" tests.

The following assay may be used in the classroom as an indirect immunological test to demonstrate the presence of HCG in urine. This test uses "artificial urine," a solution of HCG described below, circumventing the need for using real urine, which may be unavailable or unacceptable. The assay also demonstrates the interaction between an antibody and antigen, and demonstrates antibody mediated agglutination.

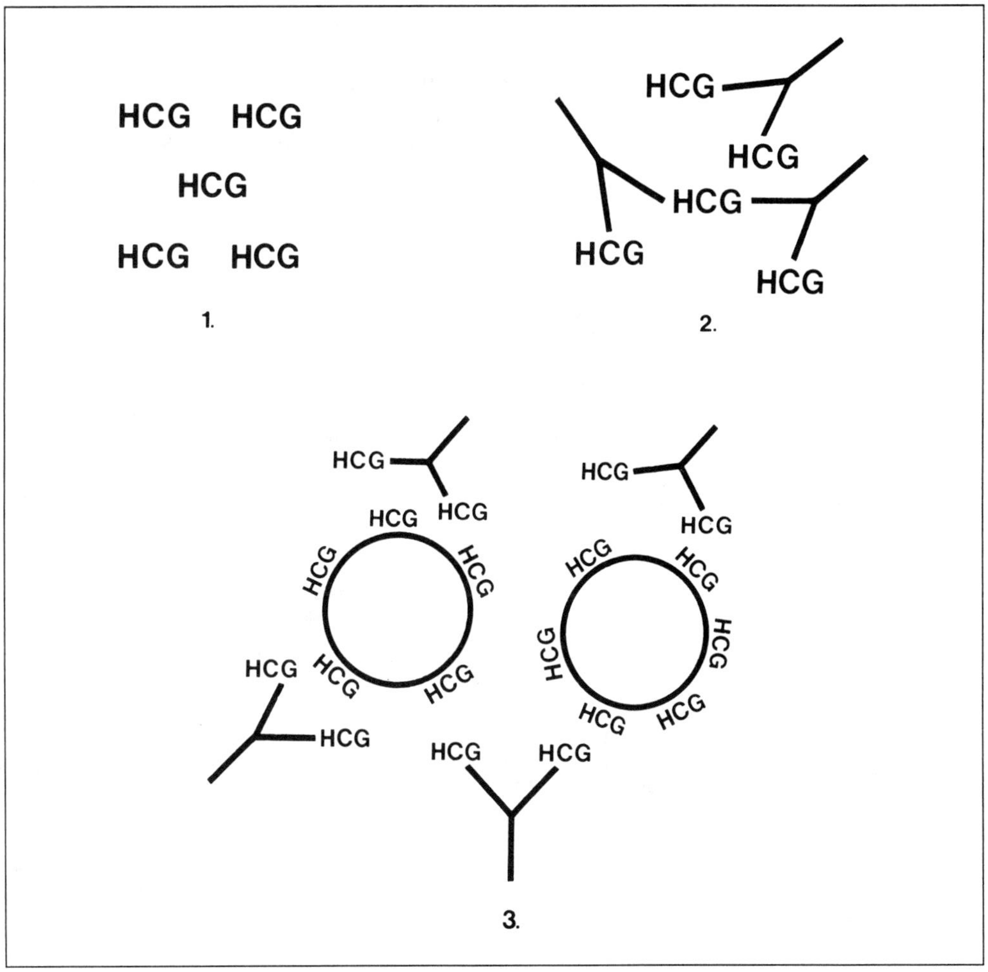

Figure 1. 1)Urine with HCG is placed on the slide. 2)Anti-HCG (Y) is added to the urine. The antibodies attach to the urine HCG. 3)HCG-coated latex particles are added. The binding sites of the anti-HCG are blocked, therefore there is no agglutination.

Simulation of Pregnancy Test

Briefly, the assay, which is done on a glass slide, involves adding HCG-urine to anti-HCG antibodies and then adding latex particles that are coated with HCG. Agglutination, seen as clumping of the latex particles, indicates that there is no detectable HCG in the urine sample. The mechanism of this is as follows. If the urine contains HCG, then this HCG will attach to the binding sites of the anti-HCG antibodies, and these antibodies will not be able to attach to the HCG-bound latex particles, and no agglutination will occur. If the urine does not contain HCG, then the anti-HCG will not be blocked and it will be able to attach to the latex particles (see Figure 1). The presence of agglutination is easily seen on a glass slide within a few minutes of adding all of the reagents together (see Figure 2).

Materials

- Human Chorionic Gonadotropin
- Anti-Human Chorionic Gonadotropin (produced in goats)
- Human Chorionic Gonadotrophin-coated Latex Suspension
- "Artificial" Urine, prepared by adding 10 ml of distilled water to the lyophilized HCG described above. The final solution will contain 250 I.U. of HCG, and 2.5 mg Mannitol/ml in 0.01 M phosphate buffer, pH 7.

Procedure

1. Place 1 drop of artificial urine, or control urine (distilled water may be used), within a specified area (circle) of a slide. The circles may be drawn on the slide with a wax pencil about 1 inch in diameter.

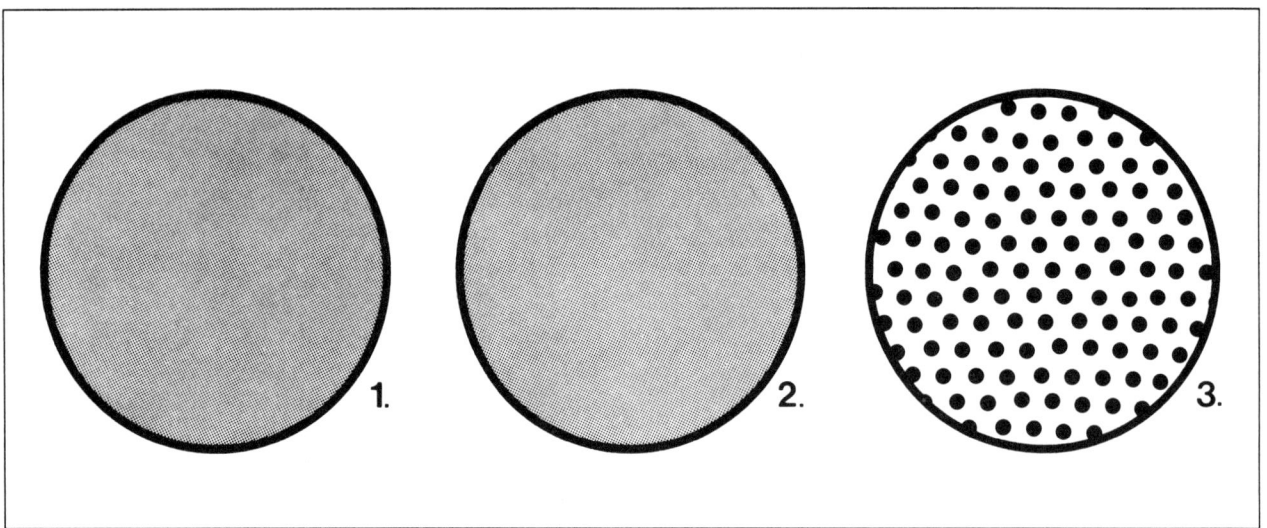

Figure 2. In this assay, "pregnant" urine was placed in Circles 1 and 2; "non-pregnant" urine was placed in Circle 3. Agglutination, seen in Circle 3 as clumping of latex particles, is seen only when urine contains no HCG.

2. Add 1 drop of anti-HCG serum to the drop of urine.
3. Add 1 drop of HCG-coated latex particles to the above 2 drops and spread all 3 drops over the entire circle with a stirrer (toothpick).
4. Rock the slide gently for 2 minutes and observe whether or not agglutination has occurred.

Suggestions

1. Since the latex particles are white, hold the slide over a black background, and hold a light source directly over the slide to facilitate the observations.
2. Have all reagents at room temperature before use.
3. Use a new stirrer for each specimen.
4. Use thoroughly cleaned slides, since traces of detergent or previous specimens adversely affect results.
5. Do not rock the slides for more than 2 minutes, since evaporation of reagents leads to errors in observations.

Observations

The assay described here is a simple, fast way of demonstrating specific antigen (HCG) to antibody interaction in the biology laboratory. The HCG test is the most common test used to detect pregnancy in humans, and therefore is very relevant to students of all ages. The assay suggests the use of "artificial" urine because of the problems inherent in handling fresh or stored "real" urine.

References

Tortora & Anagnostakos. (1984). *Principles of anatomy and physiology.* Harper and Row.

Fox. (1984). *Human physiology.* WBC.

Disease Detective: A Game Simulation of a Food Poisoning Investigation

Lisa A. Lambert, Chatham College, Pittsburgh, Pennsylvania

I often envy my colleagues in the social sciences when I hear them discuss the simulations they run in their classes. Model United Nations meetings, mock business takeovers, and studies of Middle Eastern politics all generate enormous enthusiasm among students and faculty. While simulations and games do exist for the sciences, most are directly related to community issues (like the trade-offs between energy and the environment) or clinical health scenarios (here's the patient and his symptoms: what's the diagnosis?).

I teach, among other things, a course on the biology of disease, and I wanted a simulation that would reinforce the facts I was teaching in class, that would encourage critical thinking and deductive reasoning, and that would be fun. Since I couldn't find anything that suited my needs, I decided to write my own.

The Biology of Disease course at Chatham College is a sophomore-level elective course for biology majors, and it is often taken by majors in the social sciences. One feature that distinguishes this course from other biology courses is the emphasis on epidemiological methods: studying the distribution of diseases and using this knowledge to lead to a better understanding of both the disease process and methods of controlling diseases. Throughout the course I like to use food poisoning outbreaks in a number of examples. Food poisoning is something everyone has heard of (and maybe experienced), and the different types of food poisoning (botulism, salmonellosis, etc.) can serve to illustrate a number of points such as the seasonal cycles of some diseases and differences in incubation periods, reservoirs, virulence, etc.

Plot Design

Because the subject of food poisoning plays such a prominent role in my course, it naturally came to mind first when I began to design my simulation. The real-life role of the public health official as a "disease detective" is a fascinating one, and simulating such an investigation proved to be relatively straightforward. I chose salmonellosis as the particular disease to be investigated because it's one of the most common causes of food poisoning, and, unlike botulism, its virulence or mortality rate is very low. Its typical symptoms (diarrhea, abdominal cramps and nausea) are common in many such diseases.

Scene of the Crime

So as not to upset our food service employees, I decided the initial exposure of the victims to the food poisoning microorganisms would take place at an imaginary party, held late one Friday afternoon. All participants in the simulation were told when they attended this party, whom they saw there, and what they ate.

The Victims

Twelve people participated as "victims" in this simulation, including faculty, students not enrolled in the class, and staff. Members of the class were restricted to the role of

Disease Detective

Thank you for agreeing to participate in this simulation of an investigation of food poisoning. This assignment will be handed out on _____, and will be due _____. During this two-week period, members of the Biology of Disease class will be asking you questions concerning what you ate on Friday, _____. The information below will help you answer their questions. Do NOT volunteer information (unless your script calls for it). Part of the investigative process is to learn what questions to ask. If a student fails to ask you an important question such as where you were when eating a certain meal or food, that is his or her problem!

Breakfast, lunch, dinner and snacks:
Just recite your usual meals and any snacks. Try to be consistent for each student, but if you slip up, don't panic. The students should know that memories can at times be unreliable. A little inconsistency will only add to the realism.

Party!
This is an imaginary party which took place in the new snack bar between 3:30 and 5:00, Friday afternoon. This, of course, is where the infections occurred. Try not to emphasize the party too much. For example, if a student asks you what you ate that day, you don't need to volunteer where or when you ate the foods. However, always answer direct questions fully.

You attended the party between: _____

You ate: _____

At the party, you saw the following people: _____

Again, once the student realizes you were at the party, don't volunteer who you saw there unless specifically asked.

You became ill: _____
Your symptoms were: _____
_____

Special instructions: _____
_____
_____
_____

Figure 1. Salmonellosis investigation participant script.

Disease Detective

investigator. (Other possible scenarios are outlined in the analysis section.) Each person was given a script (see Figure 1) with general and individualized information. Scripts were handed out to participants one week before the students received their assignment. This allowed time for any questions to be answered or conflicts to be resolved. While the participants had to stick to the script on particulars concerning the party, they were free to improvise on foods eaten the rest of the day. Many people really got into the spirit of things and described dinners at new restaurants in town, or picnics complete with turkey and dressing and egg salad.

Before handing out the scripts, I plotted the complete scenario and constructed a summary sheet for my own reference (Table 1). Of the 12 people participating, eight became ill, and four did not. Two of the 12 were intentional "red herrings;" one ate the infective food and did not become ill, and one did not eat the infective food, but became ill anyway, though at a much earlier time than everyone else. Attendance at the party

Table 1. Instructor summary sheet.

Participant name	Sick (Y/N)	Eating Time	Illness Onset	1	2	3	4	5	6	7	8
CLB	Y	4:15 p.m.	9:30 a.m.	-	+	+	+	+	+	+	-
JB	Y	4:15 p.m.	8:30 a.m.	-	-	+	+	-	+	-	-
MK	Y	4:30 p.m.	8:30 a.m.	-	+	+	-	+	+	+	-
MS	Y	3:30 p.m.	11:00 a.m.	-	-	+	-	+	+	-	-
JW	Y	3:30 p.m.	9:00 a.m.	+	+	+	+	-	+	+	+
LW	Y	4:00 p.m.	9:30 a.m.	+	+	+	-	-	+	+	+
DC	Y	4:00 p.m.	10:00 a.m.	-	-	+	+	+	+	-	-
KJ[1]	Y	4:30 p.m.	2:00 a.m.	-	+	-	+	+	+	+	-
KC[2]	N	4:00 p.m.	---	+	+	+	-	-	+	+	+
CH	N	3:30 p.m.	---	-	-	-	-	+	+	-	-
WB	N	4:15 p.m.	---	-	-	-	+	-	+	+	-
CM	N	3:30 p.m.	---	+	-	-	+	-	+	-	+

Food Identification Chart

Number	Food	Vulnerable/Not Vulnerable
1	Cream	Vulnerable
2	Sour cream dip	Vulnerable
3	Ambrosia salad	Vulnerable
4	Boston cream pie	Vulnerable
5	Sherbert punch	Not Vulnerable
6	Cookies	Not Vulnerable
7	Fresh vegetables	Not Vulnerable
8	Coffee	Not Vulnerable

[1]Red Herring #1: This individual did not eat the infective food (food number 3) but became ill coincidentally.
[2]Red Herring #2: This individual did eat the infective food, but ate too little to get an infective dose.

was intentionally staggered so that it was necessary to interview several people to get the names of everyone involved.

The Suspects

Every mystery has to have suspects, and in this scenario, the suspicious characters are vulnerable foods (those able to support the growth of salmonellae). These are typically moist, high-protein items such as dairy products, meat, beans and eggs. Vulnerable foods at the party included sour cream dip, ambrosia fruit salad, Boston cream pie and the cream available for coffee. Nonvulnerable foods included cookies, fresh vegetables (carrot and celery sticks), punch made with sherbert and ginger ale, and coffee.

I attempted to confuse the issue by varying my instructions to different participants. For example, some people described the ambrosia salad as "fruit salad" and others listed the ingredients. I also asked a few people to make comments such as: "That punch smelled funny, so I avoided it" or "I thought the cream was a funny color." This should not have distracted the students as they had learned that salmonellae typically cause no change in color or odor in contaminated food.

In this particular instance, the infective food was the ambrosia salad. The simulation is set up so that any combination of foods can be introduced in future years and any vulnerable food can be chosen as the culprit.

The Detectives

To prepare the students for their role as public health investigators, I lectured on the different types of food poisoning before handing out the assignment. To further complement the project, I invited an epidemiology investigator, Ms. Judy Rosenwasser from the Allegheny County Department of Health, to speak to the class after the assignment was complete. The simulation itself began Tuesday, September 20, and their report was due two weeks later. While the assignment ran, the class continued to meet twice a week as we discussed the general topic of pathogenic organisms and their life cycles.

The actual assignment was as follows:

> You are an official sent by the Department of Health to investigate reports about a possible outbreak of salmonellosis among the students, faculty and staff of Chatham College. You will have to ask the appropriate questions in order to conclude whether or not there was an outbreak of salmonellosis, and, if so, to discover how many people were affected, the attack rate, the probable infective food, etc. In addition to filling in the summary sheets on the following pages, you should write a brief report outlining your conclusions and theorizing as to how the infective food became contaminated.

Three tables show abbreviated versions of the three summary charts provided with the assignment. The first, *Individual Case Report Form* (Table 2) was to be duplicated and used for each person interviewed. The second form, *Case Histories and Summary Table* (Table

Disease Detective

3), was to be completed after the interview process. The last form, *Attack Rate Table* (Table 4), was to be used to calculate what percentage of people eating each type of food became ill. This would lead to the final identification of the infective food. As further incentive, I offered extra credit to the first student to turn in a complete, correct report.

Analysis

While my class had only 10 students, I can readily imagine problems if a class of 25 or more were allowed to pester volunteer participants over a two-week period. The next time such a simulation was scheduled, everyone would run and hide at the very sight of you. So, for a larger class I suggest that students work in pairs or small groups, or that some students be assigned as victims and others as investigators.

Another alternative design is to chose another microorganism as the causative agent of food poisoning. Staphylococcal food poisoning, for example, has a median incubation time of about three hours, and a duration of about 24 hours. Unlike salmonellae, the

---

Table 2. Individual case report form.

Name:_____ Date of Interview:_____ Sex:_____ Age:_____
Faculty/Staff/Student:_____ Date and time of onset of symptoms:_____
Description of symptoms:_____

Food and Contact 0 to 24 Hours Prior to Illness

Food Eaten	Vulnerable/Not Vulnerable	Time Eaten	Contacts at meal (if any)
_____	_____	_____	_____
_____	_____	_____	_____

Table 3. Case histories and summary table.

Person	Sick (Y/N)	Eating Time Date/Time	Sickness Onset Date/Time	Vulnerable Foods (+ or -)
_____	_____	_____	_____	_____
_____	_____	_____	_____	_____

Table 4. Attack rate table.

Vulnerable Food	Persons who did eat vulnerable food				Persons who did NOT eat vulnerable food			
	Sick	Well	Total	Rate	Sick	Well	Total	Rate
_____	_____	_____	_____	_____	_____	_____	_____	_____
_____	_____	_____	_____	_____	_____	_____	_____	_____

Staphylococcus microorganism produces an enterotoxin that is not destroyed by heat. Any food is therefore vulnerable since the chief source of the bacteria is through contamination by infected food handlers.

For students who find the idea of playing "disease detective" exciting, I recommend the book *The Disease Detectives* (Astor 1983). This book recounts some of the more famous and baffling of the cases facing epidemiological investigators from the Centers for Disease Control.

Results

Since only 10 students participated in this initial simulation, results don't statistically demonstrate the benefits resulting from participation in this exercise. However, on a more informal basis, the reactions were very positive. One or two students tried to bypass the simulation aspect by asking interviewees directly, "Are you one of the people in the food poisoning simulation?" The participants were forewarned, however, and responded to such questions with blank looks and puzzled questions. Except for a few such instances, the students took the simulation very seriously and seemed to enjoy the exercise. Several added innovations of their own, including one report that included a chart of symptoms and total hours ill. After the reports were handed in, I did hand out

---

Table 5. Student reactions to food poisoning simulation.

As compared to lectures or reading assignments, how successful was this exercise at achieving the following objectives?

	Very	Fairly	Average	Poor	Awful
Learning symptoms of salmonellosis	9	1			
Stimulating interest in epidemiology	7	2	1		
Learning about the types of food poisoning	6	2	2		
Exciting your interest in this course	6		4		
Learning the limitations of interviews	8	1	1		
Enjoying an assignment	6	1	2		

Other comments or suggestions:

1) "Good idea--have more"
2) "I cannot stress how much I really enjoyed the assignment. The assignment was a good example of applying the theories to a real-life situation. I strongly encourage its use in the near future." (Note: This student also indicated an interest in graduate work in epidemiology.)

Disease Detective a list of statements and objectives, and asked the students to anonymously rate the exercise as to how it met these objectives. The results of this very informal survey are listed in Table 5. The overall response was definitely positive.

Further Information on Games and Simulations

As mentioned at the beginning of this article, I did investigate available games and simulations for biology. I readily found some interesting board games, including one published in *The American Biology Teacher*, "The Gene Scene—A Human Genetics Game" (Mertens & Pursifull 1986). Our own library had only one reference, *Contemporary Games* (Blech 1973), that was interesting for its background material on games and simulations, but disappointing because few of the publishers of games that it listed are still in business.

A more useful text proved to be *The Guide to Simulations/Games for Education and Training* (Horn & Cleaves 1980). The first part of this book contains evaluative essays comparing games and simulations in, for example, the areas of ecology, land use and population. The second part of the book is an abstract of games arranged by discipline, complete with information on playing time and appropriate grade level. The last part of the book contains names and addresses of authors, publishers and "Other Useful Addresses," including a number of national games research laboratories.

Another more recent reference book I found was *Games and Simulations in Science Education* (Ellington et al 1981). This book contains some interesting information on the value of games and simulations while listing additional reviews of games and simulations.

References

Astor, G. (1983). *The disease detectives*. New York: New American Library.

Blech, J. (1973). *Contemporary games* (vol. 1: Directory). Detroit, MI: Gale Research Co.

Ellington, H., Addinall, E. & Percival, F. (1981). *Games and simulation in science education*. New York: Nichols Publishing Co.

Horn, R.E. & Cleaves, A. (1980). *The guide to simulations/games for education and training*. Beverly Hills, CA: Sage Publications.

Mertens, T.R. & Pursifull, J. (1986). The gene scene—A human genetics game. *The American Biology Teacher, 48*(2), 104-108.

Evolution and Ecology

Economics and Biology: An Analogy for the Presentation of the Niche Concept

Bonnie Amos, Angelo State University, San Angelo, Texas

In a first-level college biology course the niche concept is fundamental in introducing ecology. Few students at this level, particularly those in a nonmajor course, have had the opportunity to gain an appreciation for the incredible number of interactions possible among members of a biological community. This presents a difficult task for the lecturer in illustrating the richness and complexity of the niche concept. In the general biology course for nonmajors that I teach at Baylor University, I use an analogy that has been successful in communicating the multidimensions of niche strategies. The analogy is developed from a study conducted by Kangas and Risser (1979) on species packing in fast-food guilds.

I introduced the analogy by displaying a transparency that lists all of the restaurants and eating establishments in Waco, Texas. The use of this transparency and the relating of basic economic principles to ecological principles can initiate a number of discussions and introduce a number of ecological concepts. My specific objectives are to define niche and niche parameters, define competition, and introduce the concepts of resource partitioning, the competitive exclusion principle and the compression of hypothesis of ecology.

After presenting the students with the general definition of niche as the role or profession of the organism within the community, I ask them to define niche for the establishments listed. The students generally answer that the niche can be defined as providing prepared food for the people of Waco. When asked for a more specific definition the students begin to point out the differences among the restaurants and use these to define the niche for a particular restaurant. Differences commonly listed include the kinds of food served, the time required to obtain a meal, prices, the hours open, the location, ambiance, type of service (counter service versus waiter/waitress), and the presence or amount of inside seating. These factors are used to indicate the dimensions of niches and how niches can be measured. General comparisons of niche width also can be made by comparing establishments with narrow specialized niches (i.e., Super Spud or the Great Hot Dog Experience which serves only a single type of food) with those having broad generalized niches (i.e., Denny's, which is not known for any particular type of food but rather a wide variety).

Characteristics of different niches are then used to introduce and emphasize the role of competition in niche formations and community structure. To do this, the number of consumers/customers must be identified as the basic resource that is in limited supply. I use a question/answer format to help the students recognize these points. Questions such as "What must a restaurant do to stay in business?" or "Why are some establishments more successful than others?" help guide them in this direction. By the end of the discussion, the students, usually with only a little help from me in the way of questions, have established that the restaurants are dependent upon the money procured by attracting customers and their ability to attract customers is dependent upon the services they offer (niche strategies). Success is also determined, in part, by how many other establishments offer the same services (competition).

Economics and Biology

The next step is to introduce the concept of resource partitioning and to compare competitive intensities among the different restaurants. For this portion of the discussion I usually focus on fast-food restaurants (FFRs), which are well known by most college-age students. By limiting the discussion to FFRs, it is easier to establish the portion of overlap in their niches (i.e., all specialize in preparing and presenting the food in a short amount of time, similar atmosphere and types of service, comparable prices) and also point out that each clearly attempts to offer some unique product (i.e., Mexican, fish, hamburgers) or service (i.e., morning meal, open 24 hours, "We'll fix it your way"). When asked to explain the characteristic differences among the FFRs, students quickly tell me that the differences help in attracting a different segment of the customers (resources). In essence, the resources are being divided or portioned out among the different establishments (resource partitioning). At this point, I introduce ecological concepts of the competitive exclusion principle (two species cannot coexist indefinitely when they have the same niche) and the compression hypothesis (as the number of competitors increases, niche overlap and niche width decrease), drawing examples from the previous discussion along with examples from biological studies.

This analogy provides a way of introducing the ecological complexity of biological systems in a framework in which the students are comfortable and one in which they can contribute to their understanding. It also illustrates the common ground between two seemingly unrelated disciplines--economics and biology. The value of this point must not be overlooked when teaching a university required course for nonbiology majors.

References

Kangas, P.C. & Risser, P.G. (1979). Species packing in the fast-food restaurant guild. *Bulletin of the Ecological Society of America*, *60*(3), 143-148.

Daphnia--A Handy Guide for the Classroom Teacher

Betty Collins, Sagle Elementary School, Sagle, Idaho
Kevin Collins, Sandpoint Middle School, Sandpoint, Idaho

The study of arthropods in junior and high school science classes usually focuses on the dissection of crayfish and grasshoppers. While both of these are good organisms for study, working with living organisms adds a spark of student interest not found with preserved specimens.

I have successfully used *Daphnia*, a common water flea, to bring out student enthusiasm in the laboratory. I can collect them from a local seasonal pond, and they are also available from biological supply houses. A gallon jar of water holds an ample supply of specimens for about a week's study. They are large enough to see when collecting, yet small enough to allow viewing in detail under the low and medium power of a microscope.

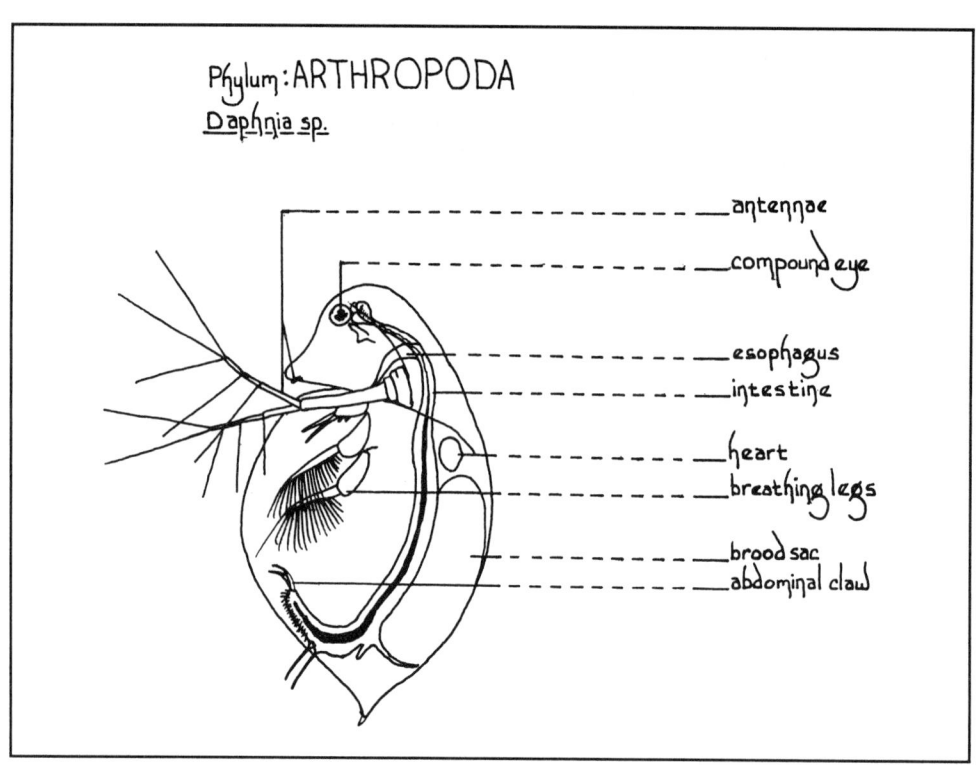

Figure 1. Adult *Daphnia*, actual size about 3 mm or less, have a transparent exoskeleton called a carapace. Heart rate, movement of food within the digestive system, movement of breathing appendages and development of young in the brood chamber are all easily recognized with a little practice. (Special thanks to Bil Sams for the drawing.)

During lab work observing *Daphnia*, my students often asked questions that were as thought-provoking as they were difficult to answer. My wife, Betty, and I began seeking the answers to some of these questions. This article is the result of our research. We offer it as a reference to other biology teachers.

Daphnia are probably the most well known group of the order Cladocera and are members of the family Daphnidae. Most of this genus live in slow moving, shallow, fresh water, such as ponds or river margins. These bodies of water may be permanent or seasonal, drying out at some point during the year (Willmoth 1967).

Daphnia--A Handy Guide

Antennae are the chief organs of locomotion. In *Daphnia*, these antennae are large and their propelling movements are interspersed with pauses during which the *Daphnia* sink with their antennae opened out to slow their downward progress (Kaestner 1970).

Daphnia respond to the chemical nature of their environment. They move toward areas of higher oxygen concentration that are more suited for their metabolism and where food supply is more abundant (Kaestner 1970). The color of *Daphnia* is an indicator of the oxygen content of water. *Daphnia* are colorless in well-aerated water and pink due to hemoglobin buildup in stagnant water (Barnes 1980). They respond to temperature changes by swimming up as the temperature lowers and swimming down as the temperature increases (Kaestner 1970).

Daphnia respond to their physical environment through the use of sensory abdominal setae and sensory hairs on the basal segment of the antennae (Hickman 1973). The movable, fused compound eye of this Cladoceran allows the animal to swim freely while orienting in any plane (Kaestner 1970).

Daphnia feed on bacteria, protozoa, nonfilamentous algae, and organic detritus, with the latter being used only when other food sources are depleted (Horton et al 1979). The Cladocera are filter feeders, using some of the four to six pairs of trunk appendages that also serve as gills taking in oxygen (Hickman 1973). The water current passes from front to back and the collected material is moved from a food groove by special setae at the basal part of the appendage (Barnes 1980). These particles are pushed forward along the midventral gully between the bases of the appendages to the mouth (Willmoth 1967). The mouth opens into an esophagus which leads to the stomach and intestine. The digestive ceca may open to the front of the intestine (Hickman 1973). Once digestion is completed, nitrogenous and other metabolic wastes are removed through the maxillary glands (Barnes 1980). Indigestible wastes are removed through the anus.

Table 1. Summary of factors influencing *Daphnia*.

Characteristic	Affected By:								
	O_2 level	Temperature	Time of Day	Condition of Individual	Scarcity of Food	Congestion	Waste Accumulation	Light	Salinity
Color of Daphnia	×								
Heart Rate		×	×	×					
Appearances of Males in Population		×			×	×	×		
Breaking Dormancy	×	×						×	×
Cyclomorphosis		×							

Daphnia--A Handy Guide

Reproduction in *Daphnia* occurs mainly through parthenogenesis. This is a process in which eggs formed in the ovary of an unfertilized female develop and form within the brood chamber. Up to 40 eggs, called summer eggs, may be formed at a time. During development these eggs go through a kind of meiosis within the nuclear membrane that gives the young a gene combination on their chromosomes different from the mother (Kaestner 1970). At birth these fully formed young are released from the brood chamber by an inward flexing of the postabdomen (Willmoth 1967).

In a stable environment, populations of *Daphnia* will be solely female and reproduce only by parthenogenesis. The life span of *Daphnia* is variable, generally lasting 30 - 60 days (Hickman 1973). An adult *Daphnia* is capable of producing a brood of eggs every two to three days (Orlans 1977).

However, in response to environmental variations (see Table 1), changes in the ovaries of the females take place that give rise to the production of eggs that will form male *Daphnia* (Kaestner 1970). Mature males mate with females forming winter eggs. These eggs are surrounded by ample amounts of yolk to maintain the egg during dormancy. Usually two winter eggs are encased in a protective structure called an ephippium. This structure is resistant to drying, freezing and passage through the digestive systems of some fish, birds and mammals, making dispersal possible. The ephippium is formed in the brood chamber and released when the female molts (Barnes 1980; Kaestner 1970). The period of dormancy within an ephippium is variable, lasting until environmental conditions are favorable to survival (see Table 1).

Seasonal changes in body form (cyclomorphosis) occur in *Daphnia*. The most notable change is the shape of the head, but the size of the compound eye and spine on the posterior end of the carapace are also affected. As temperature increases, the normally rounded head of *Daphnia* begins to elongate and become pointed and helmet-shaped (Willmoth 1967). To culture *Daphnia* for observation of cyclomorphosis, a temperature of 24.5° C produces long helmets and 7.5° C produces rounded heads (Kaestner 1970).

Cyclomorphosis appears to be an environmental adaptation, but the function of the adaptation is not presently understood. One possibility is that animals with helmets fare better against predation by fish. A study cited in Kaestner (1970) indicates large helmets affect swimming patterns in such a way that they were less likely to be preyed upon by guppies.

Because of their small size, *Daphnia* are ideal for classroom exploration. In my junior high science classes we've investigated the effects of substances such as coffee and tobacco tea made by soaking a cigarette in water for approximately two hours, as well as the effects of a rocket ride on the breathing rate and pulse of *Daphnia* using Estes® model rockets with parachute recovery. Older students might be able to explore factors affecting the sudden introduction of males in a culture or the conditions that initiate cyclomorphic changes. One thing is guaranteed however, if your zoology labs have been lacking excitement lately, the use of *Daphnia* will bring the enthusiasm back.

References

Barnes, R.D. (1980). *Invertebrate zoology*. Saunders College: Holt, Rinehart & Winston.

Hickman, C.P. (1973). *Biology of the invertebrates*. 2nd ed. St. Louis, MO: C.V. Mosby Company.

Horton, P.A., Rowan, M., Webster, K.E. & Peters, R.H. (1979). Browsing and grazing by cladoceran filter feeders. *Canadian Journal of Zoology, 57*(1), 206-212.

Kaestner, A. (1970). *Invertebrate zoology*, Vol. III. New York: Interscience Publisher (Division of John Wiley & Sons).

Orlans, F.B. (1977). *Animal care from protozoa to small mammals*. Menlo Park, CA: Addison-Wesley Publishing Company, Inc,

Palmer, E.L. & Fowler, H.S. (1975). *Fieldbook of natural history*. 2nd ed. New York: McGraw-Hill, Inc.

Venkataraman, K. (1981). Field and laboratory studies on *Daphnia carnata* King (cladocera: Daphnidae) from a seasonal tropical pond. *Hydrobiologia, 78*, 221-225.

Venkataraman, K. & Job, S.V. (1980). Effect of temperature on the development, growth and egg production on *Daphnia carnata* King (Cladocera— Daphnidae). *Hydrobiologia, 68*(3), 217-224.

Willmoth, J.H. (1967). *Biology of invertebrata*. Englewood Cliffs, NJ: Prentice Hall, Inc.

The Use of Allelopathic Interactions as a Laboratory Exercise

William H. Sharp, University of Lethbridge, Lethbridge, Canada
Teresa Dolman, University of Lethbridge, Lethbridge, Canada

Some of the most instructive laboratory exercises for introductory biology classes are those that involve the growth and interaction of living plant and animal species. However, many of these experiments involve too much time and materials to be used effectively in large classes. Short experiments that can be set up easily in one laboratory period and examined the following week are therefore very useful.

For example, allelopathic interactions between plant species can form the basis of such laboratory exercises. Allelopathy is the restriction of growth of an organism due to a chemical inhibitor produced by another organism. Such interactions can be demonstrated by testing the effect of extracts from roots, stems or leaves of one species on the germination and growth of seedlings of a second species (e.g., Groves & Anderson 1981; Stowe 1979; Wilson & Rice 1968). While such tests must be corroborated by field experiments to determine if there actually is allelopathic interaction in natural communities, they at least can indicate the possibility of an allelopathic interaction.

We have developed and tested such an experiment, which exhibits a number of characteristics of biological phenomena and allows students to gain a better understanding of biological methodology. The growth of the seedlings clearly demonstrates the inherent variability in populations, both in normal growth and in response to experimental treatment. The exercise introduces students to experimental procedure, following standard procedures used by researchers in this field. It provides an opportunity to introduce students to statistical methods, including plotting of frequency distribution curves; calculating basic statistics such as mean and standard deviation; and tests of significance, such as the chi-square and Mann-Whitney tests. Finally, the experiment can be used as the subject of a scientific report.

During local fieldwork we noticed that growth in various plant species, including crested wheatgrass (*Agropyron cristatum*), was greatly reduced under dense stands of snowberry (*Symphoricarpos occidentalis*). Snowberry often forms pure, dense stands that spread by rhizomes (Hitchcock et al 1959). When in full leaf, snowberry can develop a nearly solid canopy about 0.5 - 1 meter in height, producing low light conditions at ground level. Herbaceous plants under this canopy could be heavily shaded during their early growth. The dense stands may also reduce the availability of nutrients, but movement of soil by water could counteract such depletion since the stands tend to be less than 10 meters across. Other possible effects may be allelopathic in nature.

We conducted a preliminary test to determine if snowberry does produce chemicals that inhibit the germination and growth of crested wheatgrass. In fact, it does, and we now use the following procedure, adapted from Groves and Anderson (1981), for our introductory biology laboratory.

Leaves of snowberry are collected in the late summer and are frozen until required for the experiment. For testing, one, five and 10 gram lots (wet weight) of snowberry leaves are each ground in a Waring blender, added to 100 ml of water, and left to stand for 24 hours. Each solution is then filtered through cheesecloth, and 8 ml of each extract is added to a (different) petri dish containing four layers of paper toweling and 50 seeds of crested wheatgrass. A control sample is also prepared by the same technique, using tap water in lieu of the extract. The covered dishes are placed on supports in covered glass aquaria with a small amount of water in the bottom to reduce desiccation. The aquaria are stored in darkness at room temperature. After one week the shoot length and longest root length are measured on all the seeds in each dish, and percent germination is calculated.

Figure 1, which combines data from 12 student-run experiments, illustrated the inhibitory effect of the extract on shoot and root growth. Frequency distribution graphs (not shown) of the same data for each treatment clearly reveal a shift toward increasing positive skewness in shoot and root lengths as extract concentration increases. Because of this skewness, a nonparametric test (such as the Mann-Whitney test) rather than Student's t-test should be used to compare effects of different treatments (Zar 1974).

Of the 600 seeds sown in the 12 student experiments, the germination rates were 87, 88, 62 and 33% for the control, one-gram, five-gram, and 10-gram treatments, respectively. Chi-square analysis can be used to compare these sets of germination data.

For students who have access to personal computers, the use of a statistical analysis program provides opportunities to learn use of the computer in biological analysis and

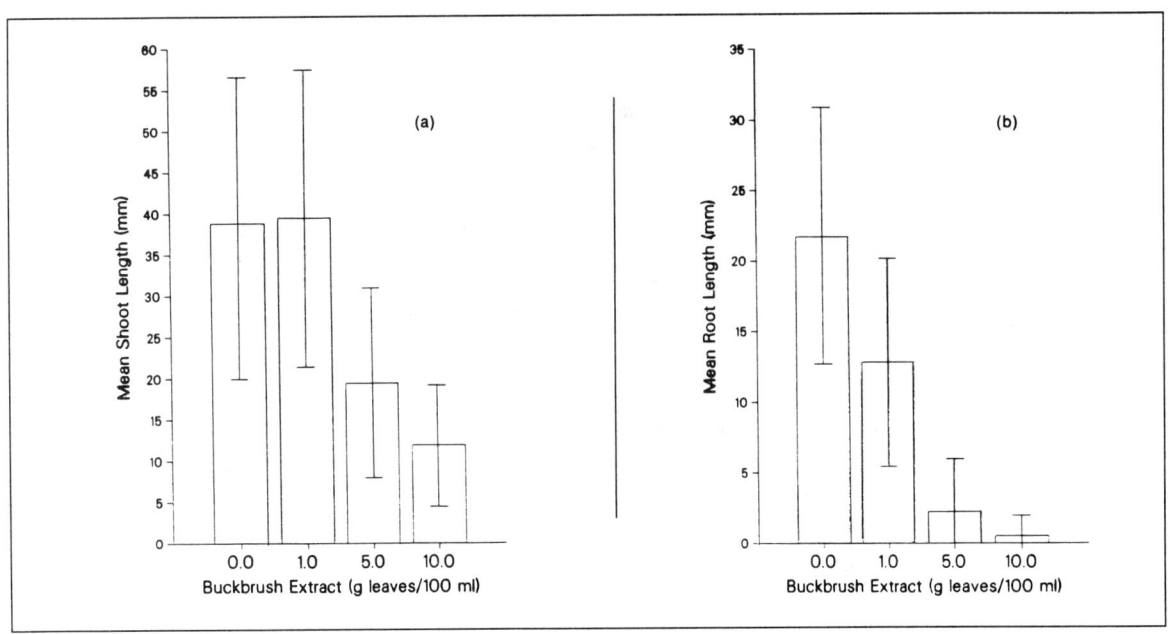

Figure 1. Effects of leaf extracts from *Symphoricarpos occidentalis* on a) shoot growth and b) root growth of *Agropyron cristatum* seedlings. Bars indicate one standard deviation. Data taken from 12 student-run experiments. N = 522, 526, 371, and 198 for the control, one-gram, five-gram, and 10-gram treatments, respectively.

Allelopathic Interactions

to utilize more sophisticated statistical techniques than would be available through other means. Statistical analysis software programs for microcomputers can be readily applied to such data. Many programs also provide plotting capabilities to display the data as bar or line graphs.

Some of the limitations of this experiment can be used as the basis for discussions. Testing for allelopathic interactions in laboratory environments does not prove that such interactions are significant in natural environments (Stowe 1979; Wilson & Rice 1968). The importance of field studies to confirm the existence of allelopathy in natural communities should be emphasized. Also, using extracts from ground green leaves does not represent normal routes of release of the inhibitory chemicals. Possible release routes include release of volatile chemicals into the air, leaf wash, decomposition of fallen leaves, or release from other parts of the plants, either alive or dead (Whittaker 1970; Stowe 1979). Incorporation of the extracts or plant parts into the soil may also affect their biological potency (Wilson & Rice 1968).

This experiment is simple, easy to set up, and requires a minimum amount of equipment. Crested wheatgrass is an introduced species that has been extensively used in agricultural and reclamation programs and has become established as a feral species in recently disturbed habitats. Seed is readily available from commercial agricultural outlets. Snowberry, *Symphoricarpos occidentalis*, is a western North American species, but members of the genus *Symphoricarpos* can be found over almost all of the continent (Gleason & Cronquist 1963; Hitchcock et al 1959).

References

Gleason, H.A. & Cronquist, A. (1963). *Manual of vascular plants of northeastern United States and Canada*. Princeton, NJ: D. Van Nostrand C., Inc.

Groves, C.R. & Anderson, J.E. (1981). Allelopathic effects of *Artemisia tridentata* leaves on germination and growth of two grass species. *American Midland Naturalist, 106*(1), 73-79.

Hitchcock, C.L., Cronquist, A., Ownbey, M. & Thompson, J.W. (1959). *Vascular plants of the Pacific Northwest. Pt. 4. Ericaceae through Campanulaceae*. Seattle, WA: University of Washington Press.

Stowe, L.G. (1979). Allelopathy and its influence on the distribution of plants in an Illinois old-field. *Journal of Ecology, 67*, 1065-1085.

Whittaker, R.H. (1970). The biochemical ecology of higher plants. In Sondheimer and Simeone (Eds.), *Chemical ecology*. New York: Academic Press.

Wilson, R.E. & Rice, E.L. (1968). Allelopathy as expressed by *Helianthus annuus* and its role in old-field succession. *Bulletin of the Torrey Botanical Club, 95*(5), 432-448.

Zar, J.H. (1974). *Biostatistical analysis*. Englewood Cliffs, NJ: Prentice-Hall.

Imbalance in Aquatic Ecosystems: A Simple Experimental Demonstration

David R. Bayer, Appleton East High School, Appleton, Wisconsin

It is most effective for high school biology teachers to get their students into the field when studying environmental concepts. Most often, however, this is a tactical impossibility because teachers often must juggle five one-hour sections that include 120 students. Practical laboratory experiences are consequently confined to the possibilities available within the school's space and schedule. By developing a realistic simulation, I have been successful at teaching students practical and important ecological concepts and enhancing their understanding of aquatic ecosystems that are out of balance due to organic pollution.

Figure 1. The experimental demonstration set up for daily class viewing.

In this experimental demonstration, the relationship between organic pollution, microbial imbalance, dissolved oxygen and fish mortality are clearly experienced. The concept of biological oxygen demand is an important one in aquatic ecosystems and is often in the news concerning local water pollution problems. The following is a description of a procedure that can give at least indirect involvement with this concept to all members of the class. It will also encourage students to do critical data analyses in order to draw conclusions.

Experimental Procedures

The demonstration setup involves three five-gallon aquaria labeled A, B and C (Figure 1). Each aquarium contains a good balance of aquatic vegetables and two small fish, one tolerant of low oxygen levels (goldfish) and one intolerant of low oxygen levels (a species of game fish—small yellow perch, bluegill or minnow). Small gamefish are often available at live bait dealers and may be mixed in with the minnow stock. All aquaria are well lighted with fluorescent lamps. Aquaria A and B then are contaminated with

Imbalance in Aquatic Ecosystems

Figure 2. Students getting estimate counts of bacteria per high power field of view after crystal violet staining.

an organic nutrient. Nutrient broth is most effective.[1] It is safe, available and, like sewage, brings on quick microbial growth. By trial and error, I have found 400 ml gives the most dynamic effect. Aquarium B is aerated with an aquarium pump connected to an air stone. Aquarium C, the control, is used to illustrate a balanced aquatic community.

Daily Observations and Data Collection

In order for the observation not to be interrupted by a weekend, the demonstration is set up on a Monday. An introduction is made at each period as a fraction of the pollutant broth is added by each class in order to dramatize the experimental variable. Data tables are passed out for all students to record daily observations. For about 20 minutes each day, four parameters are observed or measured and discussed. These parameters are water clarity, microbe populations, dissolved oxygen and the condition of the fish.

The degree of water clarity is simply observed by judgment of the eye, although more

---

[1] Nutrient broth is a high protein organic "soup" available from most biological supply companies (mix 8 grams per liter of water). Beef boullion cubes can substitute.

sophisticated means are available (Morholt et al 1966). The bold printed letters on the back face of the aquaria are a guide as to relative clarity (Figure 1). The degree of water clarity in Aquarium A will steadily deteriorate and continue that way through five days of observation. This will demonstrate the inability of stagnated water to purify itself. The aerated Aquarium B will quickly cloud up (Figure 1) within one day, and by the second day will begin to clear. By the fourth day it will appear perfectly clear. Aquarium C of course will remain unclouded.

The cloudiness of the water is relative to the abundance of microbes feeding on the nutrient contamination. Bacteria will be the first and primary microbe form, but in time an interesting succession will occur with the emergence of protozoans and rotifers. The monitoring of these microbes could be done accurately by using elaborate counting methods (Morholt et al 1966) or by simply observing a drop of water sample after crystal violet staining (Figures 2 & 3). Two students may then report to the class on the approximate number of bacteria per high power field of view, plus the variety and relative quantity of protozoans seen in the water samples. The relationship among water clarity, microbe abundance, nutrients available and dissolved oxygen is discussed each day as these factors change (Figures 4 & 5).

As microbe populations grow in Aquarium A, dissolved oxygen will drop to zero within two days because of the oxygen demand of the microbes, and will remain there throughout the demonstration. The gamefish species should die within a day, but the goldfish will remain alive, gasping at the surface, for one or two more days. In Aquarium B, oxygen level, due to constant aeration, will maintain itself through the rapid bacterial growth except for a small depression at its bacterial population peak. Then it should rise again as the nutrient pollutant is exhausted by the microbes. At that point, the microbes will die off, clearing the water. The gamefish in B may or may not die with the slight oxygen depression, but the goldfish should easily survive through the ordeal. Aquarium C's oxygen level will remain stable and the two fish species should remain healthy throughout the five days.

Concepts To Be Gained

After five days of observing and recording data on these four parameters, tentative conclusions can be drawn. By careful study

Imbalance in Aquatic Ecosystems

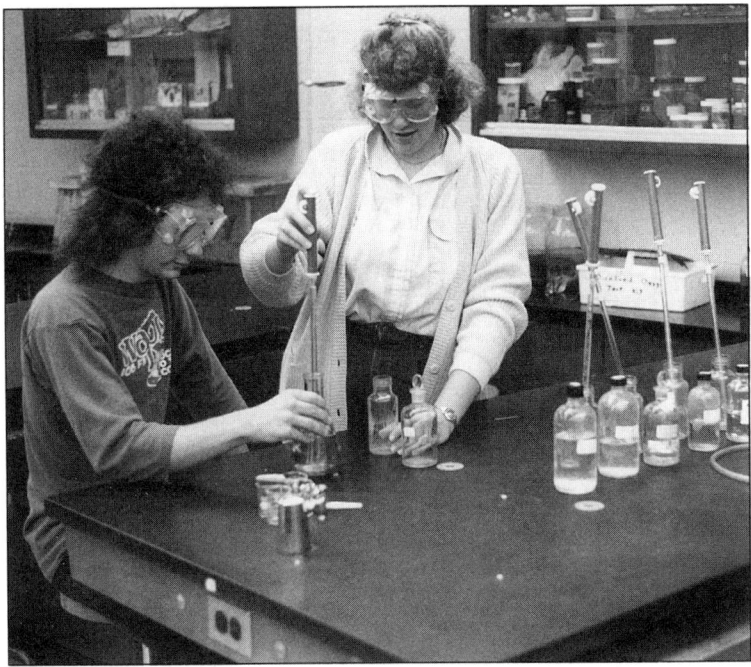

Figure 3. A pair of students daily perform the Winkler oxygen test and share their findings with the entire class.

Imbalance in Aquatic Ecosystems

of the collected data, most students should be able to begin to understand the following ecological concepts:

A. Plants maintain dissolved oxygen in a balanced aquatic ecosystem.

B. Nutrient pollutants promote a rapid increase in community decomposers. These in turn compete with other aquatic life forms for dissolved oxygen, thus creating an imbalance between decomposers and consumers.

C. On a practical level, natural and unnatural agitation or aeration of polluted waters hastens the consumption of nutrient pollutants by microbes (applications here to sewage treatment).

The Laboratory Report

In order to evaluate the students' understanding of these concepts it is expected that the laboratory report address the following questions:

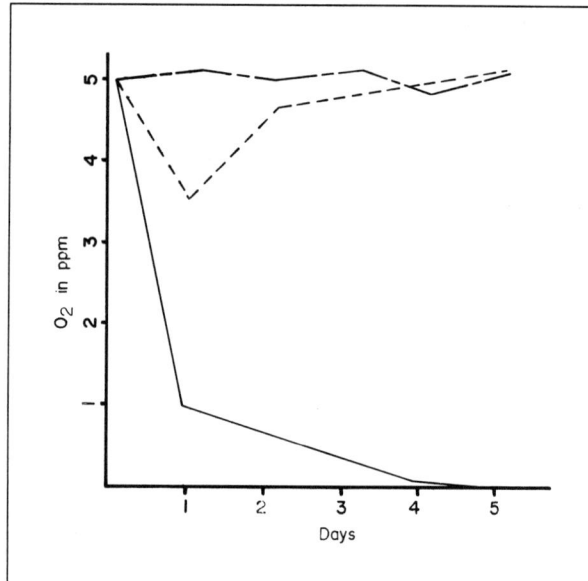

Figure 4. Daily changes are charted.

1. What direct effect did the pollutant have on the communities' decomposers?

2. Explain the relationship between water clarity and microbe populations.

3. Explain all of the day-to-day changes that happened to dissolved oxygen in each of the three aquaria.

4. Did Aquarium B or C turn clear in the time of study? If so, explain why. If not, why not?

5. Explain the death of any of the fish in any of the aquaria.

6. What is maintaining oxygen levels in the control Aquarium C?

7. How is cellular respiration by microbes related to water pollution?

8. Why do many fish die out in waters contaminated by certain industrial and municipal wastes?

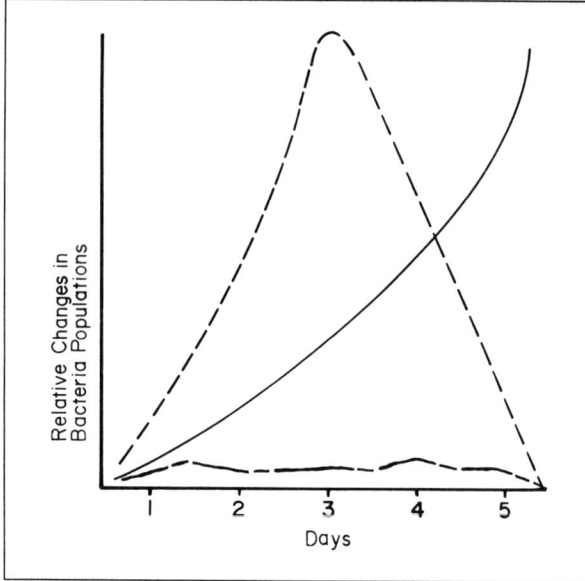

Figure 5. Bacterial population changes are compared.

9. Why are fast running rivers less likely to be affected by water pollution?

The students' degree of ability to analyze the data and answer these questions is related, of course, to their background in biological concepts. Only after dealing with the subjects of photosynthesis and respiration are meaningful interpretations possible. Furthermore, a study of bacteria and other protists, their form and growth requirements, also seem prerequisite. I have therefore found this laboratory investigation most revealing to students in a year-end unit on ecosystems.

This pertinent concept of ecosystem imbalance is difficult enough for students to appreciate without a field study. At least in this case of aquatic ecosystems, we can effectively bring it indoors where students have time to collect and analyze data that reach meaningful conclusions. This will hopefully bring them closer to an appreciation of the complexities of nature, which I believe is essential to the survival of our life support systems on this planet.

References

Morholt, E., Brandwein, P.F. & Alexander, J. (1966). *A sourcebook for the biological sciences.* New York: Harcourt Brace Jovanovich.

A Hands-On Simulation of Natural Selection in an Imaginary Organism, *Platysoma apoda*

Steve Fifield, University of Minnesota, Minneapolis, Minnesota
Bruce Fall, University of Minnesota, Minneapolis, Minnesota

Evolution is the central unifying concept in modern biology. When Theodosius Dobzhansky (1973) wrote "nothing in biology makes sense except in the light of evolution," he was describing its importance to students, not just professional biologists. To develop an appreciation of the power of evolutionary ideas, students need more than lectures on the mechanisms of evolution. We can make evolution come alive for students by teaching it as a dynamic process, rather than as a set of static definitions. Fortunately, basic evolutionary processes are relatively easy to model, and students can be active participants in the models.

Numerous laboratory simulations of evolution have been described (McComas 1991). For example, using computer simulations students can investigate the mechanisms of evolution and the effects of parameters such as population size and selection coefficients (Ortiz-Crespo 1987; Price 1993, 1985). Computer models are tremendously useful, but we feel it is worthwhile to more directly involve students as actual participants in an evolutionary process. However, some hands-on exercises may be too simplistic for undergraduates (Hinds & Amundson 1975; Kramm 1977; House 1986) or fail to incorporate fundamental processes such as sexual reproduction and Mendelian inheritance (Stebbins & Allen 1975; Bishop & Anderson 1986). This article describes the Natural Selection Exercise, which we developed from exercises similar to those described by Stebbins and Allen (1975) and Bishop and Anderson (1986). This activity, however, includes sexual reproduction and Mendelian inheritance, making it a more realistic simulation of natural selection. We use the exercise in an introductory college biology course with an enrollment of 200 - 300 students per quarter, most of whom are not biology majors.

In the Natural Selection Exercise, evolution occurs in a population of imaginary organisms, *Platysoma apoda*, represented by circular paper chips punched out of construction paper. The organisms inhabit a multi-colored poster like that shown in Figure 1. A single gene with two alleles controls body color in *P. apoda*, and another unlinked gene with two alleles controls body size. Students prey upon *P. apoda*, and their selection results in adaptive evolutionary changes in the prey population. Using the exercise, students can study the effect of predation on allele frequencies, examine the assumptions of the Hardy-Weinberg model and consider whether the "need to survive" is a guiding force in evolution.

Conducting the Exercise

We use a variety of simulations, which involve different combinations of traits, dominance and linkage relationships, and environments. In this paper we describe three simulations that differ in the environments used and the traits under selection (Table 1). Simulations I and II compare the effect of selection on body color in different environments, while leaving the dominance relationship of the alleles unchanged. Simulation III examines the effect of selection on body size in a population of a single color. The

Natural Selection Simulation

Minneapolis poster, used in Simulations I and III, is an aerial caricature of the city. The Stadium poster, used in Simulation II, is a caricature of a crowd at a University of Minnesota football game. Both posters, which are approximately 0.7 m², have complex patterns with many colors present, but only one or two dominant colors. These posters are no longer commercially available, but any poster or fabric with a suitably complex and colorful design can be used.

A group of four students is assigned a simulation and given three sets of paper chips representing the different genotypes. Since it is necessary to distinguish homozygous dominant from heterozygous individuals during the mating procedure described below, inconspicuous symbols designating the genotypes are printed on the chips. (See the appendix for more details on preparing the paper chips.) Two of the students, acting as helpers, assemble a starting population of 100 individuals consisting of 25 of each homozygote and 50 heterozygotes. The starting frequency of each allele in the population is therefore 0.50. These 100 individuals are thoroughly mixed and spread across the environment poster. The other two students in the group, acting as predators, then remove a total of 50 individuals. Each predator removes 25 individuals, selecting the first individual he or she sees and depositing it in a container before returning to the poster for the next predation event. Predation events should be as independent as possible, so predators are asked not to search the same area of the poster twice in succession.

When the predators are finished, the helpers collect and sort the survivors by genotype. We assume that the survivors mate randomly among themselves and that each pair produces four offspring before dying. The 50 survivors therefore produce 100 offspring to start the next generation. The helpers calculate allele frequencies among the survivors, use a random mating table generated from the Hardy-Weinberg equation to determine the genotype frequencies for the next generation and record their data on a form that we provide. (Copies of the handout that our students receive, including the data sheet and random mating table, are available from Bruce Fall.) The random mating table lists expected genotype frequencies among 100 offspring based on the frequencies of two alleles in a randomly mating parental population. In the Hardy-Weinberg equation ($p^2 + 2pq + q^2 = 1$), expected frequencies of the homozygous dominant, heterozygous and homozygous recessive

Table 1. Parameters of the three simulations used in the natural selection exercise.

	Simulation		
	I	II	III
Environment poster	Minneapolis	Stadium	Minneapolis
Polymorphic trait under selection	body color	body color	body size
Alleles	chestnut (A)* olive (a)	chestnut (A)* olive (a)	normal (B)** dwarf (b)

*Chestnut is completely dominant over olive and all individuals are normal in size.

**Normal is completely dominant over dwarf and all individuals are olive in color.

Natural Selection Simulation

genotypes are represented by the terms p^2, $2pq$ and q^2, respectively (Hartl 1988). Students use the random mating table, rather than manually pairing the surviving individuals, to save time and minimize procedural errors. It is important that students understand the origin and purpose of the table so that the process of assembling the next generation does not become a set of directions that are followed but not understood. An example of the steps followed when using the data sheet and random mating table is shown in Figure 2. When the helpers have assembled the second generation, the next round of predation begins. The process is repeated through four generations or until one of the alleles becomes fixed. Most groups finish the exercise in 1 - 1.5 hours.

After all groups have conducted the exercise, class mean results for each of the three simulations are distributed to the students. Student presentation of results, group discussions and individual writing assignments are used to help students analyze the results of the exercise. Bishop and Anderson (1986) have identified a number of common

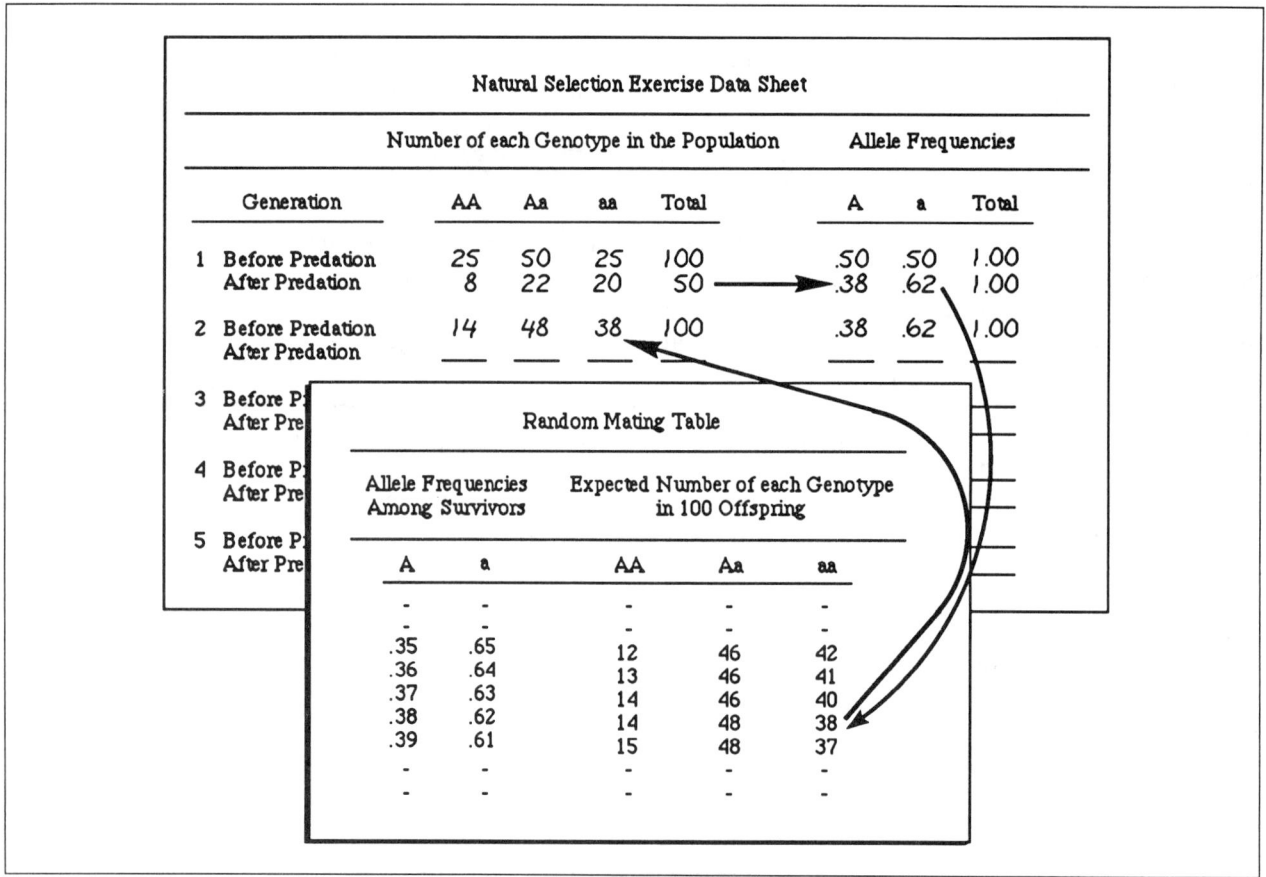

Figure 2. The data sheet shows the results of predation during the first generation. After calculating allele frequencies among the survivors, students refer to the random mating table to determine the number of each genotype in the next generation. For illustrative purposes, only a portion of the random mating table is shown.

misconceptions concerning evolution that many students must confront and correct in order to develop an adequate understanding of evolution. We have found that group discussions following the exercise help students generate interpretations of the results, identify their misconceptions and organize their thoughts for the individual writing assignment.

Results and Analysis

Effect of the Environment on Platysoma apoda

Figures 3, 4 and 5 display student data for the three simulations listed in Table 1. Students begin their analyses by interpreting the results of the three simulations based on class mean data. For example, why does the olive allele increase in frequency in the Minneapolis environment (Figure 3), while falling in frequency in the Stadium (Figure 4)? The Minneapolis poster is dominated by shades of green in which olive colored individuals are well concealed. Predators, who depend on color vision to locate their prey, are more likely to select the chestnut-colored individuals on the Minneapolis posters, leaving fewer of them to pass on chestnut alleles to the next generation. Conversely, on the Stadium poster, chestnut-colored individuals tend to blend into the maroon background and survive at higher rates than olive-colored individuals. Differential survival, and consequently reproduction, of individuals based on their phenotypes results in changes in allele frequencies in the two simulations after four generations. In both environments natural selection produces adaptive changes in the *P. apoda* populations, since the allele coding for the more successful phenotype increases in frequency. This analysis challenges students to integrate their understanding of alleles, genotypes and phenotypes, and to consider the relationship between the reproductive success of individuals and genetic change in a population.

Variation Among Predators

The mean data shown in Figures 3, 4 and 5 conceal a significant level of among-group variation in final allele frequencies. In Simulation I, for example, one group of predators drove the chestnut allele to extinction, while several other groups preyed more heavily upon the olive-colored individuals than on chestnut individuals. Students are asked to compare their group's results to the class mean results to emphasize that the reproductive success of an individual is a function of both its genes and the selection pressures it experiences. Predators exert a major selective pressure on the *P. apoda* populations, but all predators are not alike.

Is a Dominant Allele Better Than a Recessive Allele?

A common misconception held by students is that a dominant allele is inherently better than a recessive allele. This belief seems to be based on the popular connotations of the terms. To challenge this perception, we have designed two of the simulations to show the recessive allele increasing in frequency. In Simulation I the recessive olive allele

Natural Selection Simulation

increases in frequency, and in Simulation III the recessive dwarf allele rapidly increases in frequency relative to the dominant, normal-sized allele. In Simulation II the dominant chestnut allele becomes more common, not because it is dominant, but because it codes for a color that blends into the Stadium environment. This demonstrates that both dominant and recessive alleles can code for beneficial traits and that either may become common in a population. There is, of course, a difference in the rate at which dominant and recessive alleles respond to selection. Since deleterious dominant alleles are expressed in both homozygous and heterozygous individuals, they can be removed from a population more rapidly than deleterious recessive alleles that are hidden from selection in heterozygotes. This can be demonstrated by conducting two simulations using the same environment and alleles, but switching the dominance relationships of the alleles in the two simulations.

The Hardy-Weinberg Model

Students are introduced to the Hardy-Weinberg model in a lecture previous to or concurrent with the Natural Selection Exercise. They are sometimes initially unimpressed by a law that states the conditions under which no evolution will occur. Of course, the importance of the Hardy-Weinberg model lies in its identification of mechanisms that affect the genetic structure of populations: mutation, gene flow, genetic drift, nonrandom mating and selection. If none of these evolutionary mechanisms occurs, allele and genotype frequencies will not change (Hartl 1988). It is therefore possible to study the effects of these mechanisms and to track allele and genotype frequencies over many generations. Students use the principles of the Hardy-Weinberg model to calculate allele frequencies among the survivors of predation and to create the next generation of *P. apoda* using the random mating table.

With the Hardy-Weinberg model in mind, students examine the conditions of the exercise to identify all the mechanisms of evolutionary change that are present. Memorizing a list of evolutionary mechanisms and their definitions is easier than determining whether they are at work in the simulation, but searching for the mechanisms in the exercise helps students go beyond abstract definitions to actually visualizing evolution-

Table 2. Survivorship and relative fitness of the genotypes in Simulation III after the first generation (based on mean data from 19 groups).

Generation 1	*AA*	*Aa*	*aa*	Total
No. of individuals before predation	25	50	25	100
No. of individuals after predation	11.3	22.5	16.2	50.0

Survivorship: *AA* = 11.3/25 = .452; *Aa* = 22.5/50 = .450; *aa* = 16.2/25 = 648.
Relative fitness: *AA* = .452/.648 = .698; *Aa* = .450/.648 = .694; *aa* = .648/.648 = 1.0.

ary mechanisms. The only mechanism that consistently operates in the exercise is selection and, as predicted by the Hardy-Weinberg model, it results in evolution in *P. apoda* populations. Mutations do not arise during the exercise, no gene flow occurs between populations, and mating is random. Using the random mating table prevents genetic drift due to unrepresentative sampling from the gamete pool of the survivors. However, because the population size is reduced to 50 individuals every generation, some genetic drift may occur during the process of predation, contributing to variability among groups conducting the same simulation.

To explore the effects of other evolutionary mechanisms, students are asked to describe what might happen if the exercise were modified to incorporate additional mechanisms of evolutionary change. For instance, with natural selection still in place, what might be the fate of a beneficial mutant allele? In this case, students examine the likely fate of an allele that is initially present as a single copy in a gene pool of 200 alleles. Although it codes for a beneficial trait, an allele present at low frequency may be lost from the population due to chance. The effect of dominance and recessiveness on the fate of a new allele must also be considered. A dominant allele will produce the beneficial phenotype immediately, increasing the probability that the allele will spread in the population. A beneficial recessive mutant allele faces a longer period of vulnerability since it must appear in the homozygous state before natural selection can favor its increase.

Calculating Relative Fitness

The reproductive success of the genotypes in each simulation can be quantified by calculating their relative fitnesses. Relative fitness is the average contribution of a genotype to succeeding generations expressed as a fraction of the most successful genotype's reproductive success. In nature, genotypic fitness is influenced by two factors: survival and reproduction. In the natural selection exercise, all survivors produce two offspring, so fitness is equivalent to survival.

Students calculate the relative fitness of each genotype based on the class mean data from the first generation of each simulation. An example of these calculations using data from Simulation III is shown in Table 2. The fraction of each genotype surviving predation is first calculated by dividing the number of survivors by the number present before predation. These values are then expressed as relative fitness by dividing each by the survivorship of the most successful genotype. In our example (see Table 2), homozygous dominant and heterozygous individuals as a group had only about 69 percent of the reproductive success of homozygous recessive individuals due to their greater vulnerability to predation. Fitness values can be entered into population genetics computer simulations, such as EVOLVE (Price 1993), permitting students to predict the future genetic structure of their population or to reconstruct its history. Computer simulations become more relevant when one of the parameters (e.g., fitness) has been generated from real data.

Peppered Moths, Paper Chips & the Need To Survive

Natural Selection Simulation

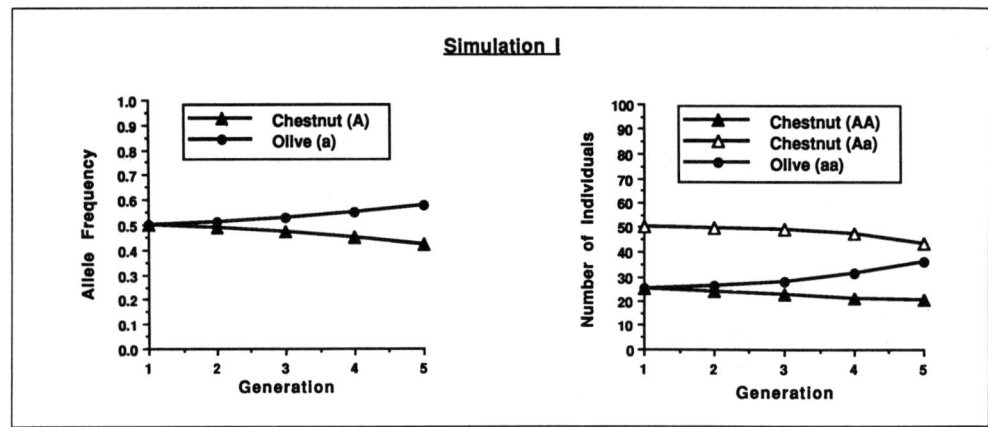

Figure 3. Change in allele (left) and genotype (right) frequencies in *P. apoda* in Simulation I (based on mean data from 23 groups).

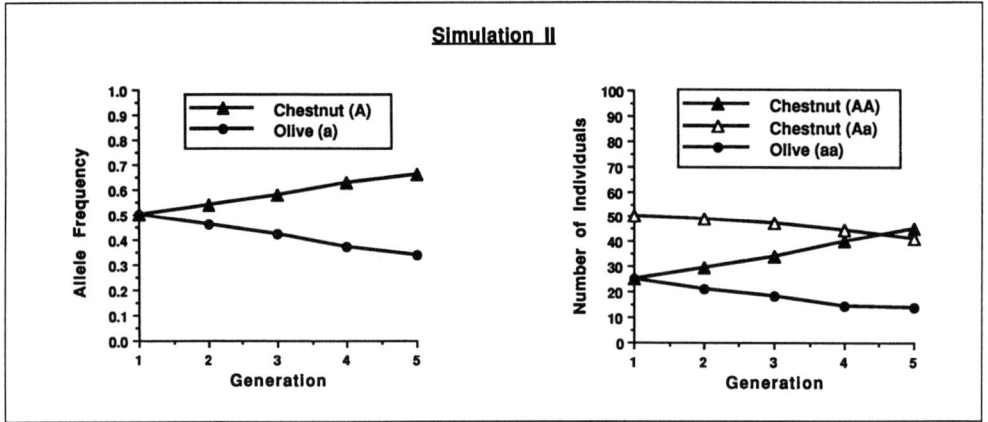

Figure 4. Change in allele (left) and genotype (right) frequencies in *P. apoda* in Simulation II (based on mean data from 17 groups).

A classic example of natural selection based on cryptic coloration is the case of peppered moths (*Biston betularia*) in England (Kettlewell 1959; Bishop & Cook 1975). Before the rise of coal-powered industry in England, most peppered moths were light-colored with scattered patches of darker pigmentation. A darker melanic form was very rare. Peppered moths are most active at night, spending their days resting on surfaces such as tree trunks and rocks. In unpolluted environments, the speckled pattern of the light-colored form blends well with the lichen-encrusted surfaces of trees and rocks, protecting the moths from predation by birds. The dark form contrasts with this background and suffers higher levels of predation. An effect of the Industrial Revolution was increased atmospheric pollution; it killed the lichens and darkened the moths' habitat, fundamentally altering the pattern of predation on moth populations in polluted areas.

Natural Selection Simulation

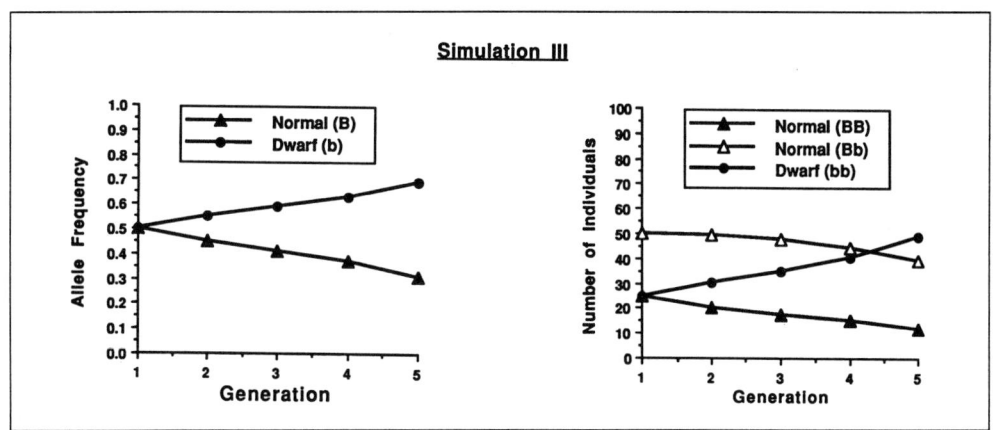

Figure 5. Change in allele (left) and genotype (right) frequencies in *P. apoda* in Simulation III (based on mean data from 19 groups).

On a darker background, the melanic form gained the survival advantage and increased in frequency in many populations. This example of industrial melanism was so dramatic that by the late 1800s the melanic form was more common than the light-colored form in some areas.

Because evolution in both *P. apoda* and peppered moths involves cryptic coloration, students compare the evolutionary mechanisms and outcomes in the two cases. When students realize that the process they participated in is analogous to events that occur in nature, they better appreciate the role that modeling can play in studying natural phenomena. A comparative approach also addresses some of the misconceptions held by students. For example, after reading about peppered moths, many students find no problem with the following statement: "The melanic form of peppered moths increased in frequency in England in the late 1800s because the moths needed to become dark in order to survive." The fallacy in this statement is that adaptation is a goal-oriented process that results in what is needed by organisms, rather than the result of natural selection on random genetic variation in a population. When asked if changes in color or size in the *P. apoda* populations occurred because the paper chips needed to survive, students are comfortable with the notion that there was no goal-oriented process at work and that pieces of paper are not known for contemplating how to survive. It is then an easier task to convince students that no vital force is at work in the evolution of peppered moths or other living things.

Conclusion

Over the years, we have modified the exercise to suit our needs, and we encourage you to tailor it to your situation. We have found it to be particularly effective to combine the hands-on nature of the exercise with the speed and flexibility of computer simulations. But no matter what method we choose, if one of our goals as instructors is to help

Natural Selection Simulation

students construct an understanding of evolution, we must give them the opportunity to explore the causes and consequences of natural selection.

Acknowledgments

We would like to thank Gordon Brown, Richard Peifer and Catherine Zabinski for helpful comments on early drafts of the manuscript.

References

Bishop, B.A. & Anderson, C.W. (1986). Evolution by natural selection: A teaching module. (Occasional Paper No. 91). East Lansing, MI: Institute for Research on Teaching, Michigan State University.

Bishop, J.A. & Cook, L.M. (1975). Moths, melanism and clean air. *Scientific American, 232*(1), 90-99.

Dobzhansky, T. (1973). Nothing in biology makes sense except in the light of evolution. *The American Biology Teacher, 35*(3), 125-129.

Hartl, D.L. (1988). *A primer of population genetics.* Sunderland, MA: Sinauer.

Hinds, D.S. & Amundson, J.C. (1975). Demonstrating natural selection. *The American Biology Teacher, 37*(1), 47-48.

House, K. (1986). Wooly worms and natural selection. *The American Biology Teacher, 48*(4), 242-245.

Kettlewell, H.B.D. (1959). Darwin's missing evidence. *Scientific American, 200*(3), 48-53.

Kramm, K.R. (1977). Demonstration of population genetics. *The American Biology Teacher, 39*(9), 558-559.

McComas, W.F. (1991). Resources for teaching evolutionary biology labs: An analysis. *The American Biology Teacher, 53*(4), 205-209.

Ortiz-Crespo, F.I. (1987). Elements of population genetics with the microcomputer. *Journal of College Science Teaching, 17*(1), 28-30.

Price, F.E. (1985). EVOLVE: A computer simulation for teaching labs on evolution. *The American Biology Teacher, 47*(1), 16-24.

Price, F.E. (1993). EVOLVE: A computer program simulating natural selection, genetic drift and gene flow. The BioQUEST Library, Academic Development Group, University of Maryland, College Park.

Stebbins, R.C. & Allen, B. (1975). Simulating evolution. *The American Biology Teacher, 37*(4), 206-211.

Appendix

Natural Selection Simulation

We use heavy construction paper and a standard 1/4 inch (6.4 mm) paper punch to cut normal-sized chips and a 3/16 inch (4.8 mm) punch for dwarf chips. Homozygous dominant and heterozygous individuals are distinguished by printing "o" and "+" on them, respectively. For consistency, homozygous recessive chips also have the "o" symbol. Using a word processor and laser printer, we generate master sheets, each filled with one of these symbols, which are aligned in staggered columns at density of about 10 symbols per cm^2. The master sheets are photocopied onto both sides of colored construction paper, and chips are punched from a stack of several sheets. The spacing and density of symbols ensures that all chips will have several symbols on each side.

To demonstrate that fitness is environment-dependent, chip colors are selected to yield opposite results on the two posters. It is easy to select two colors that have very different fitnesses on a single environment poster, but much more difficult to find two colors that give opposite results on each poster. To select paper chip colors, predation experiments are conducted on both posters using 10 colors with 15 chips of each color. From these results, a smaller number of colors are selected for paired testing.

Environmental Pollution Effects Demonstrated by Metal Adsorption in Lichens

James E. Miller, Delaware Valley College, Doylestown, Pennsylvania

Although lichens are composed of two of the lowliest life forms, a fungus and an alga, they are among the most elevated in their adaptability to environments that other organisms find intolerable.

The fungal-algal partnership forms a symbiotic team capable of survival in harsh climatic conditions and in a wide range of elevations and humidities. Lichens are among the first organisms to exploit a newly exposed piece of bedrock anywhere on this planet and to begin the soil-building and nitrogen-fixing processes necessary for subsequent invasion and succession of developing communities.

The algal component, trapped in a mesh of fungal filaments, does the photosynthetic work for the pair and often the assimilation of nitrogen. There is debate about whether the fungus really makes any positive contribution, but it is agreed that the fungal partner is responsible for anchorage to substrates and for retention of moisture. Together, however, lichens form a fascinating, richly varied and colorful flora.

The scarlet fruiting bodies of *Cladonia cristatella*, the British soldier lichen, are prominent in our drab winter landscapes, the tufts of a chartreuse *Usnea* are seen against the reddish trunks of our redwoods and incense cedars in the West, and the blue-gray mats of *Parmelia* species are found on rocks and trees all over our continent.

Two things these lichens apparently need for prolific development are a stable substrate and clean air. It can take up to two decades for a stand of lichens to develop on a fence post or a tombstone. Thus an opulent lichen growth on an old stone fence or tree trunk is an indication that the substrate has been stable for quite a long period of time.

A polluted environment can prohibit lichen growth and can kill existing populations on even the most stable and permanent of substrates. The use of salts to keep winter highways ice-free, for example, has eliminated lichen flora from treated roadsides.

The susceptibility of lichens to air pollutants is well documented. A study by Brodo (1971) follows the decline of lichens along a transect from the clean air at Montauk on the eastern tip of Long Island to the polluted air of New York City at its western end. A similar study in England (Hawksworth & Rose 1970) relates air quality to lichen flora in the rural areas and in the urban areas of Manchester and London. Excellent reviews of the effects of air pollution on lichen growth appear in a volume by Ferry et al (1973).

Materials and Methods

To perform the experiment you will need to have on your reagent shelf (and you probably do) some methylene blue and a number of metal chlorides, including a selection of mono-, di- and trivalent chlorides. $NaCl$, KCl, $CoCl_2$, $CaCl_2$, $FeCl_3$ and $AlCl_3$ will do nicely for a start. Add others as available. Either you or the students should prepare 10 mM solutions of the chlorides (about 5 ml per student group).

Metal Adsorption in Lichens

You will also need a grip-operated paper punch, a tea strainer or aquarium net, a handful of foliose lichens, the usual laboratory glassware, and a spectrophotometer for reading the results. If a spectrophotometer is unavailable, the intensity of the blue color at the end of the experiment can be assessed visually.

Obtaining the necessary lichens should not be a problem if done well in advance. So that you will know where to find them, look for foliose lichens—usually blue-gray or blue-green growths on trees or rocks—as you drive to work or walk in the woods. The fruticose (hanging and branching) lichens will not suffice, nor will the crustose lichens that cling closely to their substrates. What is needed are the foliose or leafy lichens that grow in irregular circles on trees or rocks, the kind that can be removed in sheets large enough to be cut into small circles with a paper punch.

Finding a way to incorporate lichens into the curriculum is a problem for biology teachers. A study of their taxonomy with field identification requires several weeks and in most cases is too much. A quick reference to their three major growth forms—foliose (or leafy), fruticose (or branched) and crustose (or flat crusts)—is too little, especially if included only as an interesting aside to a study of either fungi or algae.

The laboratory exercise suggested here provides a convenient and interesting experience with lichens and can be done in a period of two to three hours or in several successive shorter periods. It requires minimal preparation, which can be done by the teacher well in advance, and it forms a "hands-on" basis for meaningful discussion of algae, fungi and the lichen symbiosis.

A variation of this experiment was published in England by Gooday (1982). What is presented here is a somewhat Americanized and modified version adapted to fit a va-

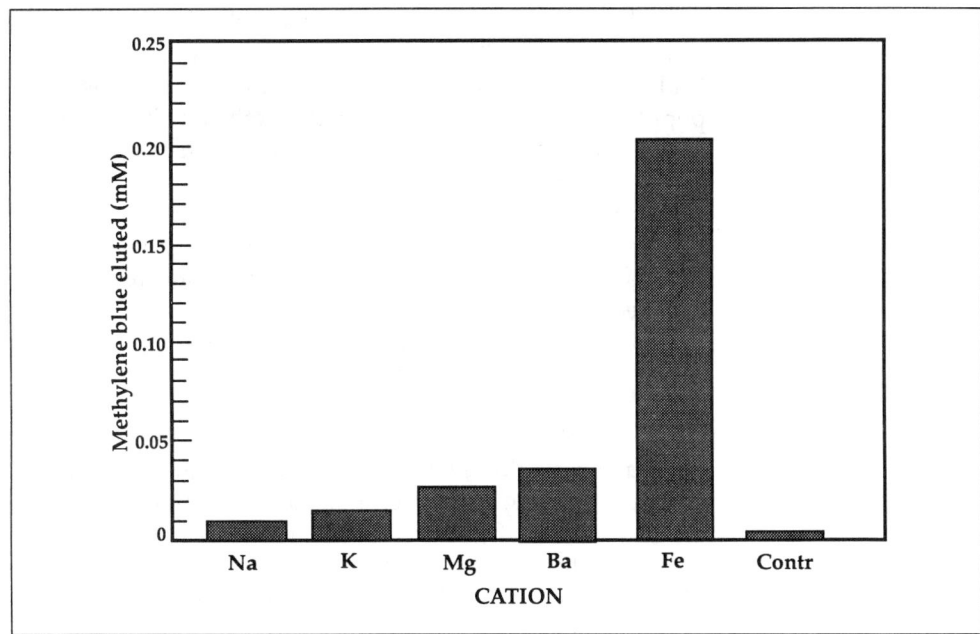

Figure 1. Methylene blue eluted by five lichen disks after one hour of incubation in various chloride solutions.

Biology Labs That Work: The Best of How-To-Do-Its

Metal Adsorption in Lichens

riety of available time slots. An attractive aspect to this experiment is that it generally gives conclusive results. In other words, it "works."

Purpose

This experiment will demonstrate in the laboratory the degree to which environmental cations are taken up by susceptible lichen organisms in the field.

In the reference by Gooday (1982) cited above, a species of *Peltigera* common in the British Isles was used. In America and Canada, *Parmelia* species are much more common. They are recognized by their blue-gray circular thalli, often several inches in diameter. If you live in an urban area where air pollution has effectively diminished lichen intensity, you can probably find some foliose lichens on stacks of cut firewood imported into the city from the hinterlands. Any species of lichen will be suitable, but be sure to collect only from areas where they are plentiful.

Prepare a solution (you will need about 100 ml) of methylene blue at a concentration of 5×10^{-4} M. If precise measurement is a problem, simply dilute some methylene blue with water until it has a color intensity just deep enough that you cannot see through it. Instruct the students to cut out disks of lichen with a grip-operated paper punch (yielding circles about the size of a shirt button). Cut enough disks so that five can be put into each chloride solution to be tested (plus five into a water control).

Drop all the lichen disks into the methylene blue solution and let them adsorb the cationic methylene blue dye for approximately half an hour, during which time you can lead a discussion of the lichen symbiosis or the effects of air pollution on lichen populations. Then wash the excess methylene blue from the disks with distilled or deionized water. A tea strainer or aquarium net is especially useful for this.

Transfer five washed disks to each 10 mM chloride solutions for one hour at room temperature, shaking them occasionally. During this interval the methylene blue cations previously adsorbed by the lichens are released as metallic cations are exchanged. This cation exchange results in the development of gradually increasing blue color to the supernatant liquid. This can be decanted from the lichen disks and evaluated at once or saved for evaluation in a future laboratory period.

If a spectrophotometer is available, the methylene blue concentrations of the supernatant liquids should be determined at a wavelength of 540 nm against a standard curve prepared by using a linear series of methylene blue concentrations ranging from zero to about 0.50 mM. If you prefer to estimate the methylene blue concentrations with the naked eye, the intensity of blue color can be rated on a scale ranging from zero to four pluses (++++), in which zero is colorless and 4+ approximates the color intensity of Windex® or Dawn®, for windows or dishes, respectively.

Results and Discussion

Since cations are among the components of polluted environments, the results of this experiment can be related to the disappearance of lichens from polluted areas. Whether the cations enter the environment from air pollutants, highway salt treatment, chemical dumping or other means, it is obvious they will have a detrimental effect upon life forms. These laboratory results represent one way to quantify those effects.

Generally, the trivalent cations are more readily exchanged with methylene blue than the divalent cations, which, in turn, are more readily exchanged than the monovalent cations. Most of the metal ions adsorbed from the natural environment are toxic to lichens, and several lichens appear to have a special affinity for iron (Anderson & Treshow 1984). Results from one classroom performance of this exercise are shown in Figure 1. Visual evaluations of the results shown here were recorded as follows: K, N(+); Mg, Ba, Co, Ca(++); Fe(+++). It is only by coincidence that the number of pluses assigned to the results happen to correlate with the valences of the cations used here.

Our results were obtained with thalli of *Parmelia caperata* collected in New Hampshire and New Jersey. Variations in results are to be expected when using different lichens, different cations and different times, but the pattern shown here will prevail.

It is very difficult for novices to identify lichen species accurately without the help of microscopy and a few chemical tests. It should therefore be emphasized that any foliose lichens will be appropriate for this experiment, but teachers may want to consult one or more of the various published keys to lichen identification. Those by Hale (1969) and by Bland (1971) are two which are especially useful for the inexperienced.

References

Anderson, F.K. & Treshow, M. (1984). In M. Treshow (Ed.), *Air pollution and plant life* (pp. 259-289). New York: John Wiley & Sons.

Bland, J.H. (1971). *Forests of Lilliput: The realm of mosses and lichens.* Englewood Cliffs, NJ: Prentice-Hall.

Brodo, I.M. (1971). Lichens and air pollution. *The Conservationist*, 26(1), 22-26.

Ferry, B.W., Baddeley, M.S. & Hawksworth, D.L. (1973). *Air pollution and lichens.* Toronto: University of Toronto Press.

Gooday, G.W. (1982). Cation adsorption by lichens. In S.B. Primrose & Wardlaw (Eds.), *Sourcebook of experiments for the teaching of microbiology* (pp. 82-84). New York: Academic Press.

Hale, M.E. (1969). *How to know the lichens.* Dubuque, IA: Wm. C. Brown Company Publishers.

Hawksworth, D.L. & Rose, F. (1970). Qualitative scale for estimating sulfur dioxide air pollution in England and Wales using epiphytic lichens. *Nature*, 227(5254), 145-148.

Using *Lemna* To Study Geometric Population Growth

Larry E. DeBuhr, Missouri Botanical Garden, St. Louis, Missouri

Philosophers of science tell us that one characteristic of a good theory is that it is fruitful; that is, a good theory leads to new ideas and new research (Root-Bernstein 1984). Likewise, a good topic in biology class should also be fruitful. It should lead beyond a simple understanding of the immediate content topics into discussions that tie together new ideas and make connections to other disciplines.

One particularly fruitful area of study in a biology class is population growth. A study of population growth provides an opportunity for developing a basic understanding of underlying ecological principles such as the j-curve, s-curve and environmental resistance, but will also lead to discussions of a social, cultural and economic nature. An understanding of population growth relates to a variety of STS issues such as the impact of increasing populations on the environment, the implications of more people on limited resources, and changing cultural reproductive practices.

A clear understanding of the environmental and social consequences of population growth depends upon understanding how populations or organisms grow and how these populations are naturally controlled. The reliance on the development of an understanding of basic concepts which can then be applied to societal problems is in keeping with the 1982 position statement of the National Science Teachers Association (NSTA). This statement said that, "...a scientifically and technologically literate person: uses science concepts, process skills, and values in making responsible everyday decisions." (NSTA 1982).

Some classroom lessons on geometric population growth use activities that model the concept. These models may be with dice (Christensen 1981) or may be computer simulations. Other lessons use microorganisms with rapid growth rates such as bacteria or yeast (Reynolds 1975).

Although these approaches are useful, they do have limitations. It is more difficult for beginning students to conceptualize the increase in number of bacteria or yeast cells when their numbers are inferred from turbidity measures or from other indirect counting techniques. Furthermore, many secondary schools do not have the facilities for the growth of micro-organisms and many teachers do not have the experience in culturing them. Students have more difficulty with measuring the growth of micro-organisms; these more complex lab procedures are difficult to do. Computer modeling systems are abstract and lack connections to the "real world." Students have more difficulty with transferring the concept from dice to natural systems; the models are not tied to student experiences.

I have been using a laboratory activity that has proven to be very successful. This activity has repeatedly given my class good results and has always worked well. It is an extremely easy activity to do, is inexpensive and requires a minimum of laboratory equipment. The lesson is appropriate for secondary level biology classes as well as for college courses.

Lemna and Geometric Population Growth

The activity uses a small floating aquatic flowering plant of the genus *Lemna* to investigate geometric population growth. *Lemna* is small enough to have a sufficiently rapid reproductive rate yet large enough to be easily seen and counted. Within three months, a definite j-curve can be obtained, and within five to six months, *Lemna* will entirely cover the surface of a 10-gallon aquarium with an estimated 10,000 individuals.

One of the more important characteristics of the activity is that the student can collect and analyze data on the population size of a real organism rather than on a model. The collection and analysis of data are important characteristics of science and ought to be components of good lab activities (Blosser 1983). Some lab activities on population growth do not include any data collection, but simply have students graph data collected by someone else or data from printed material (Newman 1988). These types of dry labs are pedagogically unsatisfying and do not reflect the nature of science.

The activity described in this paper allows for the integration of mathematics and graphing techniques, but out of a necessity to analyze and interpret data. The use of computers with graphics capabilities can also be incorporated into this lesson. Most importantly, the lab provides a concrete experience with an organism that is large enough to enable students to see individual plants and for which the population size is easy to determine. Students do not rely on abstract representations of geometric population growth.

The lesson was designed to follow a learning cycle format (Lawson 1988). An initial exploration phase involves collecting data on the population growth of *Lemna*. This phase is followed by an interpretation of the geometric curve tied into a discussion leading to the development of basic population growth concepts. The last phase helps students apply the concepts to human population growth and societal issues.

Growing *Lemna*

Lemna, commonly called duckweed, is a very small aquatic flowering plant with a reduced vegetative morphology. There are 15 species in the genus worldwide (Correll & Correll 1972). The most common species is *L. minor*. The plants are composed of 1 - 3 small leaves that are rarely larger than 3 mm in length and 2 mm in width. Each leaf bears a single root (Cronquist 1988).

Although *Lemna* produces small flowers and seeds, its rate of vegetative production is high, and it is this latter method that is the primary mode of reproduction. As individual leaves grow and enlarge, they break apart from the parent plant and form new individuals. *Lemna* can reproduce and double its number in less than five days if the growing conditions are adequate. As *Lemna* grows and reproduces, it forms a population of floating plants on the surface of the water.

A definite advantage of *Lemna* for population growth activities is the ease with which it can be cultivated. I have had very good luck growing *Lemna* in an aquarium with a hood

Lemna and Geometric Population Growth

and light. The aquarium held aged water aerated with a single air stone. I do not use a filter because I do not want a great deal of turbulence at the surface of the water. Although gravel in the bottom of the aquarium is not needed, it does add some aesthetic qualities. The light is left on at all times. Initially the water level will drop due to evaporation, but the rate of evaporation will decrease as the size of the floating mat increases. I replace the evaporated water with aged tap water.

This method of growing *Lemna* is very low cost. Most schools already have aquaria or other appropriate tanks that could be used. The setup can be aerated with a simple, inexpensive pump. *Lemna* can be field collected or purchased from a supply house. I have also seen it growing in aquaria at tropical fish shops. Once a culture is established, it can be used for years.

Another advantage of *Lemna* is that it forms a single layer of plants on the surface of the water. It is very easy to count individual plants because they are not piled on top of each other. After the surface of the aquarium is filled, *Lemna* begins to grow in a multilayered mat, and counting or estimating the population size is considerably more difficult. I have estimated a population size of more than 200,000 plants in a 10-month-old culture. The plants were growing in a layer 2-3 cm thick which covered the surface of the aquarium water.

Lab Procedure: Exploration

Because of the time involved with this activity, it is necessary to start the lab five months prior to when the lesson on population is introduced. The specific timing is not critical, however, since geometric population growth will be demonstrated after about 75 days. Do not let the time involved with this activity discourage you from trying it, as it is well worth the extra effort necessary for advanced planning.

Several days prior to the class period when the lesson is introduced to the students, set up the aquarium as described above. *Lemna* can be grown in any size aquarium, although I have found that a 10-gallon aquarium works well and gives good results within a reasonable time period. An aquarium much larger than 10 gallons will require more time for *Lemna* to reproduce and fill the aquarium.

Prep the students for the activity by telling them that they are going to begin a more lengthy activity, one that will take several months. Let them know that they will be collecting data on the population growth of *Lemna*. They will want to know what *Lemna* is, so let them see some *Lemna* plants. At this time tell them what you have done to prepare for the activity (set up the aquarium, aerator, no filter, etc.). Begin the activity by placing 10 *Lemna* plants in the aquarium. The students may want to know why you didn't use only one plant. (There is a possibility that any *Lemna* plant will die. If only one plant were used, and it died, you would no longer have a population. If 10 plants were used, and several died, the population would continue.) You might like to have the class predict the reproduction rate and doubling time of *Lemna*.

Lemna and Geometric Population Growth

At this point, it is appropriate to introduce students more fully to *Lemna*. Students should have an opportunity to study the plant under a dissecting microscope. They should be introduced to the morphology and general biology of the genus. Lessons in development or asexual vs. sexual reproduction might be considered. You might like to assign library research on the genus and have students do the background reading. Some questions to consider: What is the geographic range of *Lemna*? What ecological conditions does it grow in most often?

After a discussion of *Lemna*, pass out data collection sheets and explain how to collect and record the population size. Initially students should count the number of *Lemna* plants in the aquarium. This is fairly easy up to about 1000-1200 plants. The plants tend to cluster in small clumps and counting the plants is fairly easy. Counting the number of plants takes only a few minutes every couple of days; many students finish their counting before class starts or after class is over. In order to encourage independent work habits, I require students to work alone as they collect their data.

The timing of the counting is not crucial. Initially students will want to count every day. The students will discover that for several days, no population increase will occur, and then within one day, the population will increase significantly. These "spurts" in reproduction are to be expected and represent important lessons in growth and development.

As the lesson progresses, it will be necessary to remind students to continue counting. It is also advisable to check data sheets on a regular basis in order to monitor student progress.

During your planning for this activity, reserve a class period for a lesson on how to indirectly measure the population size of *Lemna*. After several months (around 1000-1200 plants), the number of plants becomes too large to easily count. At this point, open up discussions with the students as to how they could get an idea of the population. Try to get the students to invent a method that is appropriate. Help the class evaluate each idea as it is suggested.

One method used by students for an indirect measure of the population size is to:

1. Calculate the surface area of the aquarium.

2. Average, from 10 samples, the number of *Lemna* plants per cm^2.

3. Estimate the percentage of the surface covered by the plants.

4. Calculate the population size by multiplying 1. × 2. × 3.

The aquarium used to collect the data in Table 1 has a surface area of 1212.75 cm. From a sample size of 10, students averaged the number of plants/cm^2 at 7.3. Using this figure and the percentage of the surface area covered with plants, the population size was

Lemna and Geometric Population Growth

calculated. For example, on day 83, 30% of the surface area of the aquarium was covered with *Lemna*. This was 363.8 cm² (30% times 1212.75 cm). Therefore 363.8 cm² × 7.3 plants/cm² = 2656 plants.

Table 1 shows a set of data collected from one class using a 10-gallon aquarium. The table includes data on the population size from direct measurement by counting and also estimated data using the method described above. An examination of the data in Table 1 shows that the population growth is not perfectly geometric; the data will seldom be ideal. Students should be helped to understand that generalizations are often "perfect" representations of an "imperfect" world.

Students who count on the same day may not have the same number of population sizes on their data sheets. Although students should be taught the importance of accurate data collection, the time that students count the number of plants will vary and small errors in counting will occur. These errors will be more common as the population size increases and as students estimate the population size with indirect measures.

Ideas for other lessons have resulted from student questions or observations during the course of the activity. One student suggested that the introduction of a competitor species such as algae or some other aquatic species might alter the population growth curve. Another student noticed that a few *Lemna* growing in a fish aquarium appeared to be larger than the plants growing in a dense mat. Students were asked to design experiments to test these hypotheses, although time was no longer available during the school year to complete experiments in class. An elementary teacher enrolled in a summer class took home a jar full of *Lemna* for his fish tank and discovered that his goldfish were very fond of using *Lemna* for food. He suggested introducing a grazer as another variable in the activity.

Table 1. Population growth of *Lemna*.

Direct Measurement by Counting							Indirect Measurement by Estimation		
Day	Population Size	Day	Population Size	Day	Population Size	Day	% of Surface Covered	Population Size
1	10	18	67	32	203	83	30	2656
5	15	19	70	33	248	110	50	4427
6	17	20	71	38	318	124	75	6640
7	17	21	73	40	406	144	85	7525
9	21	24	85	44	530	157	100	8853
10	25	25	107	47	550			
12	30	26	113	54	728			
13	40	27	125	59	877			
14	42	28	146	72	1108			
17	59	29	159	74	1325			

Students will often be amazed at how rapidly the surface of the aquarium fills up near the end of the activity. The idea that the population size doubles within a certain time period is very dramatically demonstrated with this activity. One student, upon seeing the large increase in the number of *Lemna* between day 144 and and day 157, termed the increase the "big bang."

Interpretation and Introduction of Concepts

After students have finished collecting data, ask them to graph the results. The graph will be a typical geometric curve. Discuss with them the nature of the curve and what it means when a curve increases in this manner. Introduce the idea of a j-curve. Have the students calculate the doubling time (the time it takes for the population to double in size) and the population growth rates for a particular time period (final population size/initial population size for the time period). Perhaps you will want the students to predict future populations if the growth rate continues. What are the implications when population growth is "out of control"?

It is interesting for the students to discover, as shown by the data in Table 1, that the doubling time increases as the population density in the aquarium increases.

Some thought-provoking questions to discuss with students are: If organisms follow a geometric growth curve, why isn't the world hopelessly overpopulated with plants and animals? Why haven't most populations of organisms followed the j-curve? Why don't populations of natural organisms go out of control? Since this hasn't happened, some factors must be working to control the population. Have the students identify natural factors that help control population sizes and introduce the concepts of environmental resistance, limiting factors and the s-curve.

Application to Human Population Growth

An obvious extension of this lesson is into the area of human population growth. Following the discussion described above, pass out a sheet with human population figures. Table 2 contains human population sizes as reported in Christensen (1984) and by the World Resources Institute (1987). Have the students graph human population growth. This graph should stimulate discussions as to why humans do not follow the s-curve. Is environmental resistance operating with humans? What is the future for human population? What decisions must be made? What are the implications of human population growth for future resource use, for disease control or for environmental quality?

Unquestionably, the study of geometric population control should be part of every secondary and college biology curriculum. The importance of this issue to the survival of the human species cannot be stressed too much. The lesson described helps students understand the nature of geometric population growth and has direct application to a wide range of social and environmental issues.

Lemna and Geometric Population Growth

Table 2. World population.

1650	.47 billion	1940	2.3 billion	1985	4.8 billion
1750	.69	1950	2.5	2000*	6.1
1850	1.1	1960	3.0	2025*	8.2
1900	1.6	1970	3.6	2100*	10.2
1930	2.1	1980	4.4		

*Projected

References

Blosser, P. (1983). The role of the laboratory in science teaching. *School Science and Mathematics, 83*(2), 165-169.

Christensen, J.W. (1984). *Global science: Energy, resources, environment laboratory manual.* Dubuque, IA: Kendall/Hunt Publishing Co.

Correll, D.S. & Correll, H.B. (1972). *Aquatic and wetland plants of southwestern United States.* Washington, DC: Environmental Protection Agency.

Cronquist, A. (1988). *The evolution and classification of flowering plants.* Bronx, NY: The New York Botanical Garden.

Lawson, A.E. (1988). A better way to teach biology. *The American Biology Teacher, 50*(5), 266-278.

National Science Teachers Association. (1984). *Science-Technology-Society: Science education for the 1980s.* Washington, DC: National Science Teachers Association.

Newman, B. (1988). *Biology research activities.* Annapolis, MD: Alpha Publishing Company.

Reynolds, W.W. (1975). *Laboratory manual for man, nature and society.* Dubuque, IA: Wm. C. Brown Company Publishers.

Root-Bernstein, R. (1984). On defining a scientific theory: Creationism considered. In A. Montagu (Ed.), *Science and creationism.* New York: Oxford University Press.

World Resources Institute. (1987). *World resources 1987.* New York: Basic Books.

Introducing Students to Population Genetics and the Hardy-Weinberg Principle

Thomas R. Mertens, Ball State University, Muncie, Indiana

Since 1978 I have been teaching an introductory course in human genetics for in-service secondary school biology teachers. Generally, the teachers have been enthusiastic about the course and readily see why they should master basic human and medical genetics content and should teach genetics principles using human examples. One exception to this generalization has been population genetics and the Hardy-Weinberg equilibrium concept. To make a case for this subject matter, I have had to develop a rationale meaningful to my clientele. The purpose of this article is to communicate this rationale to a wider audience and to suggest some simple strategies for introducing students to the Hardy-Weinberg (H-W) principle.

Why the Objections?

Strong resistance occasionally may be encountered when you try to teach population genetics concepts and the H-W equilibrium. Among the objections to studying these topics might be the following:

1. The content is of no practical value and will therefore not be used, so why learn it?

2. The content is never taught at the secondary school level, so why should the teacher know it?

3. Secondary school students are not ready for the level of abstraction embodied in the Hardy-Weinberg principle, nor are they capable of dealing with the quantitative aspects of applying it. Why should I (as a teacher) be concerned about it?

Unquestionably (and unfortunately) many biology students and some biology teachers have the perception that biology is a nonquantitative science, and many students are threatened by the simple algebra and the quantitative aspects of introductory population genetics. Biology teachers cannot remedy all of society's aversion to mathematics and quantitative approaches to analysis, but perhaps by using some simple concrete examples the quantitative and abstract can be made meaningful to students. Specific suggestions follow.

Why Teach Population Genetics Principles?

Perhaps by considering the questions, "Why should I teach the Hardy-Weinberg equilibrium and other basic principles of population genetics?" and "How does knowing the H-W principle help students understand human genetics?", we can develop a rationale to help students see meaning in these topics. If you have ever been asked questions such as the ones that follow, you can begin to see why understanding population genetics might be useful:

Population Genetics

1. "If type O blood is a recessive trait, shouldn't O be the most rare of the ABO blood groups?"

2. "I'm confused! How can O be the most common of the blood types if it is a recessive trait?"

3. "If Huntington's disease is a dominant trait, shouldn't three-fourths of the population have Huntington's while one-fourth have the normal phenotype?"

4. "Shouldn't recessive traits gradually be 'swamped out' so they disappear from the population?"

5. "How do scientists estimate the frequencies of heterozygous carriers of certain recessive traits in human populations?"

Answering such questions becomes much easier with some understanding of population genetics. Even if the student gains an understanding only at an intuitive and nonquantitative level, reasonable answers to such questions can be formulated.

Some of the concepts that must be conveyed to students can be summarized briefly:

1. *Populations differ from families.* When dealing with a single gene trait controlled by two alleles, A and a, you will encounter the following possible matings among the two parents in specific families: $AA \times AA$, $AA \times Aa$, $Aa \times Aa$, $AA \times aa$, $Aa \times aa$ or $aa \times aa$. Each of these matings leads to specific expectations relative to the genotypes and phenotypes of the offspring. A population, on the other hand, consists of the sum total of all of these possible matings, and the genetic structure of the population will be determined by the frequency of the different matings occurring in it. If all matings were between Aa and Aa individuals, and if there were complete dominance, then the next generation would consist of $3A$- and $1aa$. Thus, while the classical Mendelian inheritance that all biology students study explains familial inheritance, gene behavior in entire populations is not quite so simply analyzed.

2. *Frequencies of alleles determine frequencies of genotypes in a population.* Ultimately, the student must be helped to understand that the frequencies of AA, Aa and aa in a population are determined by the frequencies of the two alleles A and a in that population. What students find difficult to grasp, as evidenced by several of the five questions cited above, is that dominance vs. recessiveness is not a factor in determining the frequency of AA, Aa and aa in the population. Thus, you should not expect to encounter a 3:1 ratio in a population (unless the frequency of A = frequency of a = 0.5). Recessive phenotypes (such as blood type O) may be quite common if the allele for the recessive trait is present in high frequency in the population. Even when the dominant phenotype (A-) is more abundant than the recessive (aa), the recessive allele may be more abundant than the dominant allele. To grasp such ideas almost

necessitates studying simple numerical examples and coming to grips with the H-W equation.

3. *Understanding population genetics principles is fundamental to understanding Basic Concepts of Organic Evolution.* At its most fundamental conceptual level, evolution is a change in allele frequencies. If it can be shown that in either time or space the frequencies of the alleles of a gene are changing, then you can conclude that the population is evolving. Although the H-W equation describes an equilibrium status where allele frequencies remain constant generation after generation, it is the starting point for considering the forces that can cause changes in allele frequencies and thus produce evolution. A classic example of changing allele frequencies involves industrial melanism in the peppered moth, genus *Biston*. This historically documented example of evolution by natural selection is cited in many genetics and general biology textbooks (see, for example, Gardner et al. 1991).

4. *Practical applications of Hardy-Weinberg can be demonstrated.* For example, we can relate population genetics and the H-W principle to some of the societal issues we face today: Will neonatal screening for phenylketonuria (PKU) followed by treatment with a low phenylalanine diet result in an increase in the frequency of the PKU allele in future generations? Will prenatal screening for Tay-Sachs disease followed by terminating Tay-Sachs pregnancies result in an increased frequency of surviving heterozygotes and thus, ultimately, an increase in Tay-Sachs? Answering such questions is extremely complicated, but having an understanding of the H-W principle will help in reaching meaningful conclusions.

Applications of the Hardy-Weinberg principle in medical genetic counseling can be used to illustrate to students that there is, indeed, a practical use that can be made of the H-W equation. For example, if an individual were the normal sibling of a homozygous recessive PKU, and if both parents were normal (i.e., non-phenylketonuric, but necessarily heterozygous for the PKU allele), that person might well be concerned with the risk of having a PKU child. Such a normal sibling of a PKU has a two-thirds probability of being heterozygous. If he/she is heterozygous and happens to select a heterozygous mate, then the risk of having a child with PKU is 1-in-4 for each pregnancy. The question then becomes, "What is the probability that in the general population an individual will be heterozygous for the recessive allele causing PKU?" This question can be answered if we are willing to assume that the mate is selected from a population that mates at random with respect to the trait in question and is in an H-W equilibrium. For example, if in a certain population it can be shown that one in 10,000 babies born has PKU, then using the H-W equilibrium it can be calculated that the frequency of heterozygous carriers in that population is approximately 1 in 50. (How the frequency of 1 in 50 was derived will be discussed in the section, From Concrete to Abstract.) You can then answer the question of the concerned individual: The person in question who has a sibling with PKU has a two-thirds probability of being a carrier; the mate selected from the general population has a 1-in-50 probability of being a carrier; two carriers have a one-fourth probability of producing an affected child in each pregnancy. The probabil-

Population Genetics

ity that these three events will occur simultaneously is then as follows:

$$2/3 \times 1/50 \times 1/4 = 2/600 = 1/300.$$

While this may seem like a relatively low probability, it is considerably higher than that encountered in the general population where the risk is $1/50 \times 1/50 \times 1/4 = 1/10{,}000$. On the basis of such information, a couple can then make a more informed decision about whether to risk producing a child who might have PKU.

Specific Instructional Problems

Allele Frequency

Many students are threatened by the abstract and may be helped in understanding genetics principles if teachers begin with concrete examples. For example, I have frequently been asked, "What do you mean by allele frequency?" One approach to this problem is to place in a container a known number of objects ("pop-beads" work well) of one color and a known number of a second color: 14 red and 6 yellow, for example. If you then ask the question, "What is the frequency of yellow pop-beads in the container?", some students will quickly count the beads and calculate that 6 of the 20 are yellow, so the frequency of yellow = 6/20 = 3/10 = 0.3 = 30%. Similarly, the frequency of red beads = 14/20 = 7/10 = 0.7 = 70%. Some may even note that 0.7 + 0.3 = 1 = 100%, thus accounting for all the beads in the container.

With this as a beginning, turn to the population genetics question, "What is the frequency of the PKU allele in a population of 200 consisting of 60 *PP*, 120 *Pp* and 20 *pp* individuals? The answer to this question is analogous to the following concrete example: Place in a container 6 pairs of 2 red pop-beads, 12 pairs of 1 red joined to 1 yellow, and 2 pairs of 2 yellow beads. Now ask, "What is the frequency of yellow (or red) beads in this container?" Again, with a concrete example before them, students can calculate that there are 20 pairs of beads or a total of 40 pairs of which 24 are red and 16 are yellow. Thus, the frequency of red = 24/40 = 6/10 and the frequency of yellow = 16/40 = 4/10. Having done this concrete example, students can return to the population of 200 people, and, using a similar approach, calculate the frequencies of the *P* and *p* alleles as follows:

$$\text{The frequency of } P = \frac{(60 \times 2) + 120}{400} = \frac{240}{400} = \frac{6}{10}$$

$$\text{The frequency of } p = \frac{(20 \times 2) + 120}{400} = \frac{160}{400} = \frac{4}{10}$$

Another effective concrete approach is as follows: Have the students count off and find out how many are present in the room; suppose there are 30. Next have them hold both hands in the air to represent the two genes (one from each parent) that they possess for a given trait. Now, have 10 students keep both hands wide open, have 16 open one hand and clench one in a fist, and have four clench both hands. Then ask the students to calculate the frequency of open (or clenched) hands in the room.

Population Genetics

Thus, if there are 30 students and a total of 60 hands:

$$\text{The frequency of open hands} = \frac{(10 \times 2) + 16}{60} = \frac{36}{60} = \frac{6}{10}$$

$$\text{The frequency of clenched fists} = \frac{(4 \times 2) + 16}{60} = \frac{24}{60} = \frac{4}{10}$$

This concrete example is analogous to a population consisting of 10 *AA*, 16 *Aa* and 4 *AA* individuals; students quickly grasp this analogy and thus the concept of allele frequency. They also are led to the conclusion that the sum of the frequencies of the alleles of a given gene in a population equals $6/10 + 4/10 = 10/10 = 1$ or 100%.

Random Mating & the H-W Equilibrium

To illustrate the concepts of random mating and the H-W equilibrium, one can use the following concrete example: Make up a population of 100 paired pop-beads consisting of 81 pairs of two red beads (R-R), 18 pairs of one red joined to one yellow (R-Y), and one pair of two yellow beads (Y-Y). This, of course, is analogous to a population of 81 *AA*, 18 *Aa* and 1 *aa* individuals; each pair of beads (e.g., R-R, R-Y or Y-Y) represents one organism in a population of 100. Thoroughly mix the paired beads in a container. Then have a student reach into the container (without looking) and withdraw one pair of beads, record the results (e.g., R-R), replace the pair in the container, thoroughly mix the beads, and then withdraw a second pair (e.g., R-Y). This is analogous to the random mating of two individuals, in this case *AA* and *Aa*. The two pairs withdrawn could be R-R and R-R, R-R and R-Y, R-Y and R-Y, R-Y and Y-Y, R-R and Y-Y, or Y-Y and Y-Y.

This may be an appropriate time to encourage students to speculate about which combinations are most and least probable. Students will, no doubt, grasp intuitively that the R-R with R-R is most probable and the Y-Y with Y-Y least probable. If appropriate, you may wish to consider the probability that various "mating" combinations will occur. For example, the probability that R-R will be paired with R-R = $81/100 \times 81/100 = 6,561/10,000$ and Y-Y with Y-Y = $1/100 \times 1/100 = 1/10,000$ (see Figure 1).

Furthermore, you could use the population of 100 paired pop-beads to illustrate what effect random mating will have on the makeup of the next generation. Using the procedure just described, students should randomly withdraw a pair of pop-beads and record the frequency of each combination withdrawn. After withdrawing a pair and

Population Genetics

recording the results, the paired beads must be replaced in the container and the beads thoroughly mixed before the next withdrawal is made. After making many such withdrawals (and subsequent replacements), the students can determine, within the limits of sampling errors, the frequency of the various combinations of paired beads.

The frequencies of all the different combinations of paired beads can be predicted on a checkerboard (see Figure 1) and the results summarized as shown in Table 1.

The results obtained by the students randomly withdrawing (and subsequently replacing) the bead pairs should approximate the frequencies in Table 1. The reason for multiplying by 2 in lines 2, 4 and 5 can readily be perceived by studying Figure 1, which shows that there are 2 squares in the checkerboard for each of these combinations.

If we now regard these paired beads as genotypes (R-R = *AA*, R-Y = *Aa* and Y-Y = *aa*), we can predict the results obtained from the various combinations listed in Table 1. For example, R-R × R-R could produce only one kind of "offspring," R-R, because each "parent" can contribute only one R to an "offspring" (remember meiosis). Since there are 6561 such "matings" out of 10,000 possible, 6561 R-R offspring will be produced, assuming each mating produces only one offspring. Similarly, R-R and R-Y will produce 1458 R-R offspring and 1458 R-Y, while R-Y × R-Y will produce 81 R-R, 162 R-Y and 81 Y-Y. Table 2 summarizes the results of all possible matings.

		Second Bead Pair		
		$\frac{81}{100}$ R-R	$\frac{18}{100}$ R-Y	$\frac{1}{100}$ Y-Y
First Bead Pair	$\frac{81}{100}$ R-R	$\frac{81}{100} \times \frac{81}{100} = \frac{6561}{10,000}$ R-R & R-R	$\frac{81}{100} \times \frac{18}{100} = \frac{1458}{10,000}$ R-R & R-Y	$\frac{81}{100} \times \frac{1}{100} = \frac{81}{10,000}$ R-R & Y-Y
	$\frac{18}{100}$ R-Y	$\frac{18}{100} \times \frac{81}{100} = \frac{1458}{10,000}$ R-Y & R-R	$\frac{18}{100} \times \frac{18}{100} = \frac{324}{10,000}$ R-Y & R-Y	$\frac{18}{100} \times \frac{1}{100} = \frac{18}{10,000}$ R-Y & Y-Y
	$\frac{1}{100}$ Y-Y	$\frac{1}{100} \times \frac{81}{100} = \frac{81}{10,000}$ Y-Y & R-R	$\frac{1}{100} \times \frac{18}{100} = \frac{18}{10,000}$ Y-Y & R-Y	$\frac{1}{100} \times \frac{1}{100} = \frac{1}{10,000}$ Y-Y & Y-Y

Figure 1. Various combinations of bead pairs that could be withdrawn from a container containing 81 R-R, 18 R-Y and 1 Y-Y pairs of beads. Probabilities for each combination have been calculated.

Population Genetics

Note that of the 10,000 "offspring" produced from these matings, 8100 are R-R, 1800 R-Y and 100 Y-Y; i.e. the same frequencies—81/100 R-R, 18/100 R-Y and 1/100 Y-Y as in the initial population. Similarly, the frequencies of the two "alleles" R and Y have stayed the same:

$$\frac{(81 \times 2) + 18}{200} = \frac{180}{200} = \frac{9}{10} = 0.9 \text{ for R and}$$

$$\frac{(1 \times 2) + 18}{200} = \frac{20}{200} = \frac{1}{10} = 0.1 \text{ for Y}$$

Thus, in a random mating population an equilibrium is maintained with the frequencies of the alleles (R and Y) and the frequencies of the genotypes (R-R, R-Y, Y-Y) remaining constant generation after generation. This is what is meant, then, by an H-W equilibrium. Note that we have made these assumptions in this model: Random mating occurs and all genotypes are equally viable and equally fit reproductively. That is, mates are selected without regard to their genetic constitution (R-R, R-Y or Y-Y) and all genotypes are equally likely to survive and reproduce.

Thus, beginning with the concrete model, using pop-beads, the entire H-W equilibrium concept can be developed. Once students have gained an understanding of this concrete model, the instructor can move forward with additional examples, applying the H-W equilibrium to real data from human populations or the populations of other species.

From Concrete to Abstract

The concrete example just completed allows one to generalize. If the frequency of the R "allele" is 9/10 and that of the Y "allele" is 1/10, then you can calculate the frequencies of the various "zygotic" combinations as shown in Table 3. Thus the frequencies of the two alleles determine the frequencies of the three zygotic combinations, using the "multiplication rule" as shown in the Punnett Square:

81/100 R-R + 18/100 R-Y + 1/100 Y-Y

To generalize fully using more conventional gene symbols, let p = frequency of the A allele and q = the frequency of the a allele; p = q = 1. Now calculate a general statement of the H-W equilibrium as shown in Table 4. This Punnett Square can be summarized as $p^2 + 2pq + q^2 = 1$, in which p^2 = the frequency of AA individuals, $2pq$ = the

Table 1. Summary of results from Figure 1.

1. R-R & R-R:	81/100 × 81/100	=	6561/10,000
2. R-R & R-Y:	(81/100 × 18/100)2	=	2916/10,000
3. R-Y & R-Y:	18/100 × 18/100	=	324/10,000
4. R-R & Y-Y:	(81/100 × 1/100)2	=	162/10,000
5. R-Y & Y-Y:	(18/100 × 1/100)2	=	36/10,000
6. Y-Y & Y-Y:	1/100 × 1/100	=	1/10,000

$$\text{Total} = \frac{10{,}000}{10{,}000} = 1 = 100\%$$

Table 2. Results of all possible matings.

	OFFSPRING		
	R-R	R-Y	Y-Y
1. R-R × R-R	6561		
2. R-R × R-Y	1458	1458	
3. R-Y × R-Y	81	162	81
4. R-R × Y-Y		162	
5. R-Y × Y-Y		18	18
6. Y-Y × Y-Y			1
	8100	1800	100

Biology Labs That Work: The Best of How-To-Do-Its

Population Genetics

frequency of *Aa* and q^2 = the frequency of *aa*, thereby accounting for 100% of the population.

With this generalized statement we can consider such applications as the one described earlier where we know from epidemiological data that in a certain population 1 in 10,000 babies born has PKU. Thus,

$$q^2 = 1/10{,}000$$

$$q = \sqrt{1/10{,}000} = 1/100$$

Since $p + q = 1$, it follows that $p = 99/100$. We can then estimate the frequency of heterozygotes in the population by substituting numerical values for p and q in the H-W equilibrium statement.

$$2pq = \text{frequency of } Aa$$
$$2pq = 2(99/100)(1/100)$$
$$= 198/10{,}000 \approx 200/10{,}000 = 1/50$$

Numerous other examples can be found in genetics textbooks; for example, we know that in the black population approximately 1-in-400 babies born has sickle-cell anemia, from which we can calculate that about 1-in-10 blacks is heterozygous for sickle cell. Since we know that in the U.S. Jewish population perhaps 1-in-3600 babies born has Tay-Sachs disease, can you now calculate the frequency of heterozygotes in this population? In addition, the H-W equilibrium concept can be extended to X-linked genes (e.g., classical hemophilia or red-green color blindness) or the cases involving three alleles (e.g., ABO blood groups).

One can also address the forces that can bring about changes in the frequencies of the two alleles *A* and *a*. These forces include mutation, selection, random genetic drift (Hammersmith & Mertens 1990) and migration. Intuitively one can grasp that if *A* mutates to *a* more frequently than vice versa, the effect would be to increase the frequency of the *a* allele in a population. On the other hand, if each *a* allele lost by the death of an *aa* individual is replaced by an *a* allele arising by mutation, an equilibrium condition will be maintained. While such concepts can be grasped intuitively, the mathematically more sophisticated may wish to explore more quantitative evidence to support these conclusions.

Table 3. Frequencies of the three zygotic combinations.

Eggs	Sperm 9/10 R	Sperm 1/10 Y
9/10 R	81/100 R-R	9/100 R-Y
1/10 Y	9/100 R-Y	1/100 Y-Y

Table 4. Frequencies of the three zygotic combinations using the H-W equilibrium statement.

Eggs	Sperm p A	Sperm q A
pA	p^2 AA	pq Aa
qa	pq Aa	q^2 aa

Summary

In my experience, the most basic aspects of population genetics and the Hardy-Weinberg principle can be made meaningful and useful to both students and teachers. The key is to make use of concrete approaches and to derive the H-W equilibrium statement from the concrete examples. Using practical examples showing how the H-W equation can be applied will also do much to convince both secondary school teachers and their students that population genetics can be more than an exercise in high school algebra.

References

Cummings, M.R. (1991). *Human heredity* (2nd ed.). St. Paul, MN: West Publishing Co.

Gardner, E.J., Simmons, M.J. & Snustad, D.P. (1991). *Principles of genetics* (8th ed.). New York: John Wiley and Sons.

Hammersmith, R.L. & Mertens, T.R. (1990). Teaching the concept of genetic drift using a simulation. *The American Biology Teacher*, 52(8), 497-499.

Mange, A.P. & Mange, E.J. (1990). *Genetics: Human aspects* (2nd ed.). Sunderland, MA: Sinauer Associates.

Mettler, L.E., Gregg, T.G. & Schaffer, H.E. (1988). *Population genetics and evolution* (2nd ed.). Englewood Cliffs, NJ: Prentice Hall.

Thompson, J.S. & Thompson, M.W. (1986). *Genetics in medicine* (4th ed.). Philadelphia, PA: W.B. Saunders.

General Techniques

Preparing and Diluting Solutions: An Exercise for Courses in Biology Teaching Methods

Robert Tatina, Dakota Wesleyan University, Mitchell, South Dakota

Undergraduates who are planning to teach biology in secondary schools must be proficient in the laboratory skills that the current high school biology curricula demand. These skills include the care of living organisms, aseptic techniques, use of instrumentation (pH meters, spectrophotometers, etc.), and preparation and dilution of solutions, to name a few. Although many of these skills are learned and practiced in biology courses, the preparation of solutions generally is not. Furthermore, while many of the older editions of sourcebooks written for biology teachers (Klinkman 1970; Morholt et al 1966; Schwab 1963) contain sections dealing with the preparation of solutions, the newer editions (Behringer 1973; Mayer 1978) do not. Because of the need for this skill, its possible neglect in the training of potential biology teachers and its absence from the newer sourcebooks, I have included it in my biology teaching methods course. Even the biology majors with chemistry minors who take the course find the review beneficial.

In this paper, I describe an exercise designed to assess student proficiency in the preparation and dilution of aqueous solutions. In the exercise, students master the general objective of being able to prepare solutions of any specified volume and concentration. As students meet this objective, they also learn to apply several concepts of concentration (i.e., percentage solutions including weight/weight, weight/volume, and volume/volume molarity), the rule for dilution (concentration × volume = concentration × volume), which are needed to prepare serial dilutions, and the proper use of such laboratory equipment as balances, volumetric flasks, pipets, pipeters and spectrophotometers.

I introduce the exercise with a review of terminology, definitions and calculations, followed by several demonstrations of the preparation of aqueous solutions. During the demonstrations, I stress safety procedures (i.e., proper labeling of solutions, no oral pipeting, etc.). Students are then assigned several written problems to test their mastery of the concepts and calculations. Finally, the solutions to these problems are discussed before the students are assigned specific concentrations and volumes of aqueous solutions to prepare.

Because it is difficult to observe each student perform all of the steps involved in preparing a solution, I check only the end results, and do so by comparing the student-prepared solution to an instructor-prepared standard solution using a spectrophotometer. Although this comparative check does not point out where students have made errors in their calculations or techniques, it rapidly separates those who are proficient from those who need additional help and practice. It also leads naturally into a consideration of the uses of a spectrophotometer.

A specific example of the laboratory portion of this exercise follows. The students are directed to prepare 100 ml of a 0.00050 M potassium permanganate solution. Because the students are limited to the use of quadruple beam balances (which weigh to the nearest 0.01 g), they must solve the additional problem of preparing a solution at least

10 times more concentrated than is required. Thus, they must dilute their solutions, applying the rule for dilution. Their final solutions are checked in a spectrophotometer set at 530 nm and compared for accuracy to an instructor-prepared 0.00050 M potassium permanganate solution.

I chose potassium permanganate for use in this exercise because it is inexpensive, easily handled and may be used by biology teachers [for example, in the treatment of "red legs" in frogs (Behringer 1973)]. Furthermore, potassium permanganate has a high molar absorptivity and thus exhibits strong optical absorption even at relatively low concentrations (0.0005 M to 0.00005 M). Consequently, using potassium permanganate allows for the preparation of solutions with little waste of solute. Nevertheless, students should be warned that potassium permanganate is a strong oxidizing agent and a poison and, as such, should be handled with care and respect.

Although the procedure I have described is more time consuming than having students solve only hypothetical problems, solution preparation is a closer approximation to what they will be doing when they are teaching on their own.

References

Behringer, M.P. (1973). *Techniques and materials in biology.* New York: McGraw-Hill Book Co.

Klinkman, E. (Ed.). (1970). *Biology teachers' handbook* (2nd ed.). New York: John Wiley and Sons, Inc.

Mayer, W.V. (Ed.). (1978). *Biology teachers' handbook* (3rd ed.). New York: John Wiley and Sons, Inc.

Morholt, E., Brandwein, P.F. & Joseph, A. (1966). *A sourcebook for the biological sciences* (2nd ed.). New York: Harcourt, Brace and World, Inc.

Schwab, J.J. (Ed.). (1963). *Biology teachers' handbook.* New York: John Wiley and Sons, Inc.

Simple Principles of Data Analysis

Julia H. Cothron, Mathematics and Science Center, Richmond, Virginia
Ronald N. Giese, College of William and Mary, Williamsburg, Virginia
Richard J. Rezba, Virginia Commonwealth University, Richmond, Virginia

Developing students' ability to analyze and interpret data is a valued goal of science education and a key to students becoming scientifically literate. Techniques proposed for the teaching of data analysis range from the inadequate strategy of having students answer a series of direct questions about data to teaching students how to mathematically analyze and describe their personal data in paragraph form. With the exception of Biological Sciences Curriculum Study's *Interaction of Experiments and Ideas,* science textbooks do not discuss the simple descriptive statistics necessary for data analysis. Middle and high school students can learn to use these basic descriptive statistics on data generated in class or in independent projects. Measures of central tendency (i.e., the most typical value in a set of data) and variation (i.e., spread within the data), are the most fundamental and can be taught in 3-5 class periods.

Basic Experimental Terms

A first level of data analysis is being able to determine which data describe the independent variable, the dependent variable, the constants, the control and repeated trials. A scenario such as the following provides a basis for this level of analysis. It also serves both as a quick review and mind-set for further analysis.

> *Mary investigated the effect of different concentrations of Chemical X on the growth of tomato plants. Mary hypothesized that if higher concentrations of Chemical X were added, the plants would exhibit poorer growth. She grew four flats of tomato plants (10 plants/flat) for 15 days. She then applied Chemical X as follows: Flat A, 0% chemical; Flat B, 10% chemical; Flat C, 20% chemical; Flat D, 30% chemical. The plants received the same amount of sunlight and water each day. At the end of 30 days, Mary recorded the height of the plants (in centimeters) and the color of the leaves (green, yellow green, yellow, or brown).*

The concentration of Chemical X is the *independent variable* which was purposefully changed by the experimenter. The color of leaves and height of the plants were the *dependent variables* that responded to the change. *Constants* are the factors that are held the same for all samples, e.g., amount of sunlight and water. The group that received no treatment (0% chemical X) is the *control group* and is used to assess the effect of any unforeseen variable. Each of the plants receiving fertilizer represents a *repeated trial* for that value of the independent variable. Collectively repeated trials serve to warrant increased confidence in the finding, i.e., that the results do not occur merely by chance. Some students may have difficulty with these concepts. Provide such students with several scenarios similar to the above and have them practice identifying examples of the various parts of an experiment. If you have students write abstracts of their research this year, they will make good practice examples for your classes next year.

Quantitative vs. Qualitative Data   Data Analysis

Discuss the two types of data collected in Mary's experiment described on page 194. Distinguish between the two major types of data: quantitative and qualitative.

Quantitative data

Data that are based upon measurements using a scale with equal intervals are quantitative. Examples are the height of tomato plants in centimeters, the mass of rabbits in kilograms, the temperature of water in degrees Celsius.

Qualitative data

Qualitative data are also observations but do not use a scale with equal intervals. Examples include the gender (male/female) of an organism and the color of plant leaves (brown/yellow/yellow-green/green). Such categories and the operational definitions of their boundaries may be developed by the experimenter from a review of the literature or after collecting data.

Have students describe the kinds of data reported in the above experimental scenario and then classify this data as quantitative or qualitative. Qualitative data must be mathematically analyzed differently than quantitative data.

General Overview of Data Analysis

Introduce students to the idea that there are two important ways to describe a set of data. For quantitative and qualitative data, different names and mathematical methods are used to calculate values that describe the data's central value and the variation within the data. Table 1 may be used as an advance organizer for special terms that will be introduced over several days.

Table 1. General overview of data analysis.

Two Major Questions	*Quantitative Data* *(Measurements)*	*Qualitative Data* *(Observations)*
What is the central or most typical value of the data?	Mean	Mode
What is the variation or spread in the data?	Range	Frequency distribution

Biology Labs That Work: The Best of How-To-Do-Its

Data Analysis

Table 2. Raw data: Height of plants (cm).

	Concentration of Chemical X		
0%	10%	20%	30%
15	18	12	6
14	20	10	8
13	14	14	5
15	20	10	4
15	18	8	4
17	19	8	5
18	18	10	8
12	18	10	7
19	17	11	8
15	19	12	5

Table 3. Format for data table.

	Concentration of Chemical X			
Descriptive Information	0%	10%	20%	30%
Mean				
Range				
Number				

Central value is the most typical or central value of the data set. Two common values used to describe central value are the mean for quantitative data and the mode for qualitative data.

Variation describes the spread within a data set. Two common ways to describe variation are range for quantitative data and frequency for qualitative data.

Constructing a Data Chart for Quantitative Data

Provide students with a set of raw data or measurements that were collected by Mary when she conducted her experiment to determine the effect of various concentrations of Chemical X on the height of tomato plants (Table 2).

Quantitative data analysis

Lead students into the division of a data chart by having them draw a rectangle divided into four sections. Label each section with a concentration of chemical (e.g., 0, 10, 20, 30, 70). Write the independent variable, concentration of Chemical X, across the top of the rectangle. Review the two major ways to mathematically depict a set of data by describing the central value and the variation within the data set. Ask the students to examine Table 1 and to identify the terms that are used to describe the central value (mean), and the variation (range) of quantitative data. Expand the rectangle to incorporate a column of descriptive information which includes these terms and the number of measurements made (Table 3).

Data Analysis

Provide students with definitions of the terms and examples of how to calculate:

Mean: arithmetic average of sum of the individual values divided by the number of cases.

Example: $\text{Mean} = \dfrac{\text{Sum of all values}}{\text{number of cases}} = \dfrac{5+6+4+5+7+3}{6} = \dfrac{30}{6} = 5$

Minimum: the smallest value of a variable.
Maximum: the largest value of a variable.
Range: the value obtained when the minimum value is subtracted from the maximum value. It is the difference between the smallest and largest value observed.

Example:
7 6 5 5 4 3; Minimum = 3; Maximum = 7; Range = 4
15 13 12 11 9 9; Minimum = 9; Maximum = 15; Range = 6
10 10 8 7 5 4; Minimum = 4; Maximum= 10; Range = 6

Using Mary's raw data, calculate the mean, range and number for each concentration of Chemical X. Introduce the concept of number (N) as the total of cases, samples or individuals tested in each trial. Ask the students, "Would the experiment be stronger, weaker or no different if Mary had 50 seeds per tray rather than 10?" Generally the larger the N, the more confidence you can place in your results (e.g., 75 out of 100 is more convincing than 7 out of 10). Communication is helped by including the minimum and maximum values in the data chart (Table 4).

Help the students form an appropriate title that includes a statement of the effect of the independent variable (concentration of Chemical X) on the specific dependent variable (height of tomato plants).

Qualitative Data Analysis

Provide students with a set of raw data collected by Mary when she determined the effect of various concentrations of Chemical X on the color of tomato plant leaves (Table 5). Review the major ways to describe a set of data; describe the central value and the variation within the qualitative data. Ask the students to examine Table 1 and tell you the special terms that are used to describe the central value, or mode, and the variation, or frequency distribution, of qualitative data. Draw the appropriate data chart. Provide

Biology Labs That Work: The Best of How-To-Do-Its

Data Analysis

Table 4. Effect of various concentration of Chemical X on the height of tomato plants (cm).

Descriptive Information	Concentration of Chemical X			
	0%	10%	20%	30%
Mean	15.3 cm	18.1 cm	10.5 cm	6.0 cm
Range	7 cm	6 cm	6 cm	4 cm
Minimum	12 cm	14 cm	8 cm	4cm
Maximum	19 cm	20 cm	14 cm	8cm
Number	10 plants	10 plants	10 plants	10 plants

students with the terms, modes and frequency value, their definitions and examples of how to calculate them.

Mode: the most frequently occurring value. In the case of a tie, cite both values.

Example:
Consider the color data for 0% concentration in Table 5. The mode would be green since there are eight greens and two yellow-greens. Now look at the data for 30% concentration. There are five yellows and five browns. The data are bimodal. Both values are reported (e.g., mode is yellow-brown).

Frequency distribution: the number of cases that fall into each category of the variable.

Table 5. Raw data: Color of leaves.

0%	10%	20%	30%
G	G	YG	Y
G	G	G	Y
YG	G	YG	B
G	G	YG	B
G	G	Y	Y
G	G	Y	B
G	G	Y	B
YG	G	YG	B
G	G	YG	Y
G	G	G	Y

KEY: G = green, YG = yellow-green, Y = yellow, B = brown

Again consider the color data for 0% concentration in Table 5. The frequency distribution would be reported as green, 8; yellow-green, 2; and N = 10.

Using Mary's raw data, calculate the mode and frequency distribution of the color for each concentration of Chemical X and check your results in Table 6. Help the students form an appropriate title by stating the effect of the independent variable (concentration of Chemical X) on the specific dependent variable (color of tomato plant leaves).

Extensions for Older Students

With older or more experienced students, subdivisions of quantitative data (interval versus ratio) and qualitative data

Data Analysis

Table 6. The Effect of various concentrations of Chemical X on the color of tomato plant leaves.

Descriptive Information	Concentration of Chemical X			
	0%	10%	20%	30%
Mode	Green	Green	Yellow-green	Yellow-brown
Frequency distribution	G: 8 YG: 2 Y: 0 B: 0	G: 10 YG: 0 Y: 0 B: 0	G: 2 YG: 5 Y: 3 B: 0	G: 0 YG: 0 Y: 5 B: 5
Number	10 plants	10 plants	10 plants	10 plants

Table 7. General overview of data analysis.

Type of Information	Quantitative Data (Measurements)	Qualitative Data (Observations)	
		Nominal (Categories)[1]	Ordinal (Ranked)[1]
What is the most typical or central value?	Mean	Mode	Median[2]
What is the variation or spread?	Range	Frequency distribution	Frequency distribution

[1]The subdivisions of qualitative data (nominal, ordinal) and standard deviation are introduced only to older or more advanced students.
[2]Standard deviation.

(nominal versus ordinal) may be introduced. The appropriateness of various measures of central tendency—mean, median, mode—for each type of data can be discussed (see Table 7). More precise procedures of describing variation including variance and standard deviation can be incorporated. For additional information on these descriptive techniques, see *Interaction of Experiments and Ideas*.

Evaluation of Effectiveness

Since 1982, the techniques described above have been used with students, grades 6 through 12, in the Hanover County Public Schools (Virginia). Initially, the techniques were used primarily with students completing independent research projects for the Virginia Junior Academy of Science (VJAS). At the 1981 and 1982 meetings, 43 students, or an average of 22 students per year, participated in VJAS. These students, who were mentored by two teachers, won 11 awards. From 1983 to 1990, 709 students or an average of 89 students per year, participated in VJAS. These students were mentored by

Data Analysis

about 20 teachers annually. Quality research was maintained with students receiving approximately 289 VJAS awards, including special awards for best research papers in specific categories, scholarships, and designation as Virginia's representatives to the American Junior Academy of Science.

In 1990, Hanover's emphasis changed from utilization of these skills in independent research projects to integration within the curriculum. A continuum of process skills was developed for all students, grades 6-12, which included the use of statistics to analyze data. Existing textbook laboratory activities were modified or new investigations written to emphasize data analysis, graphical displays, and written or oral reports of experimental findings. Although Hanover's emphasis changed in 1990, the division has remained a major supporter of VJAS and the newly created Metro Richmond Science Fair. From 1991 to 1993, 314 students or an average of 84 students per year participated in these events winning 98 awards. Approximately 50 teachers are involved in teaching the 6-12 science program.

In 1987 Cothron, Giese and Rezba teamed to incorporate successful elements of the Hanover program into a series of publications and pre-service/in-service programs. From 1988 to 1992 the team taught four three-credit and 17 one-credit courses on "Experimental Design and Analysis" across the Commonwealth of Virginia. Funding was provided by grants from the Federal Title II Program through the State Council of Higher Education for Virginia, Virginia Power Company, Appalachian Power Company, and the Science Service, Commonwealth of Virginia. Course participants showed significant gains on pre-post tests keyed to course objectives, modification of instructional strategies in the desired directions, and increased numbers of students participating in science competitions. Dissemination continues today though local funding, national conferences, and publications.

References

Biological Sciences Curriculum Study. (1983). *Interaction of experiments and ideas.* Englewood Cliffs, NJ: Prentice-Hall.

Cothron, J.H., Giese, R.N. & Rezba, R.J. (1993). *Students and research: Practical strategies for science classrooms and competitions* (2nd ed.). Dubuque, IA: Kendall/Hunt Publishing Co.

Cothron, J.H., Giese, R.N. & Rezba, R.J. (In press). *Science projects by the hundreds: Experimenting at home and school.* Menlo Park, CA: Addison-Wesley Publishing Co.

About the Authors

Bonnie Amos, Ph.D., is Head and Associate Professor in the Department of Biology, Angelo State University, San Angelo, TX 76909.

David R. Bayer teaches at Appleton East High School, 2121 Emmers Dr., Appleton, WI 54915.

Barton L. Bergquist, Ph.D., is Professor of Biology and Assistant Dean at the University of Northern Iowa, Cedar Falls, IA 50614.

Donald E. Brott, Ph.D., was a science instructor at Fulton-Montgomery Community College, Johnstown, NY. His current address is unknown.

Roy B. Clariana, Ph.D., is a principal training/evaluation specialist at the EG&G Rocky Flats Plant in Golden, CO. His address is 1490 S. Elizabeth St., Denver, CO 80210.

Paul E. Clifford, Ph.D., is a lecturer at School of Biology and Biochemistry, The Queen's University of Belfast, Belfast BT9 7BL Northern Ireland.

Betty Collins teaches at Sagle Elementary School, 1022 Sagle Rd., Sagle, ID 83860.

Kevin Collins is a science educator at Sandpoint Middle School, 310 S. Division, Sandpoint, ID 83864.

Julia H. Cothron, Ed.D., is Executive Director of the Mathematics and Science Center, 2401 Hartman St., Richmond, VA 23224.

Larry E. DeBuhr, Ph.D., is Director of Education at the Missouri Botanical Garden, P.O. Box 299, St. Louis, MO 63166.

Teresa Dolman is an academic assistant, University of Lethbridge, Lethbridge, AB, Canada T1K 3M4.

Donald S. Emmeluth, Ph.D., is Professor of Science, Fulton-Montgomery Community College, 2805 State Highway 67, Johnstown, NY 12095.

Bruce Fall is an associate education specialist and assistant to the director of the General Biology Program at the University of Minnesota, P180 Kolthoff Hall, Minneapolis, MN 55455.

Steve Fifield is a doctoral student in the Department of Curriculum and Instruction, University of Minnesota, P180 Kolthoff Hall, Minneapolis, MN 55455.

Jerry Foote, Ph.D., is Professor Emeritus, University of Wisconsin-Eau Claire, Eau Claire, WI 54701.

Pat J. Friedrichsen is a biology teacher at Lincoln High School, Lincoln, NE 68510.

Douglas J.C. Friend, Ph.D., is Professor of Botany at the University of Hawaii, 3190 Maile Way, Honolulu, HI 96822.

Ronald N. Giese, Ed.D., is a professor in the School of Education, College of William and Mary, Williamsburg, VA 23187.

Alan L. Gillen, Ed.D., is a biology instructor at Tomball High School/Tomball College, 13705 Sandy Lane, Tomball, TX 77375.

Patrick Guilfoile, Ph.D., teaches at Whitehead Institute for Biomedical Research, 9 Cambridge Center, Cambridge, MA 02142.

David R. Hershey, Ph.D., is Adjunct Professor in the Department of Biology/Horticulture at Prince George's Community College, 301 Largo Road, Largo, MD 20772-2199.

Stuart W. Hughes, Ed.D., taught biology for 31 years at Central High School in Philadelphia. He now teaches part-time in the biology department of La Salle University, Philadelphia, PA.

Anne A. Kamrin is a science teacher at The Baldwin School, Bryn Mawr, PA 19010.

Carole B. Knickerbocker is a compliance specialist with the Pennsylvania Department of Environmental Resources, 55 North Lane, Suite 6010, Lee Park, Conshonocken, PA 19428.

Lisa A. Lambert, Ph.D., is Assistant Professor of Biology at Chatham College, Pittsburgh, PA 15232.

Joan S. LaVan is a science teacher at The Baldwin School, Bryn Mawr, PA 19010.

John E. Lennox, Ph.D., is Associate Professor of Microbiology at the Altoona Campus of The Pennsylvania State University, 3000 Ivyside Park, Altoona, PA 16603.

Lois T. Mayo is Science Department Head and biology teacher at Pius X High School, Lincoln, NE 68510.

Thomas R. Mertens, Ph.D., is Professor Emeritus of Biology, Ball State University, 2506 Johnson Road, Muncie, IN 47304.

James E. Miller, Ph.D., is Professor of Biology, Delaware Valley College, Doylestown, PA 18914.

Randy Moore, Ph.D., is Dean of the Buchtel College of Arts and Sciences at the University of Akron, Akron, OH 44325.

Robert Moss, Ph.D., is Assistant Professor of Biology at Yeshiva University, 500 West 185th St., New York, NY 10033.

Steven B. Oppenheimer, Ph.D., is Director, Center for Cancer and Developmental Biology, The California State University, Northridge, CA 91330-8303.

David S. Ostrovsky, Ph.D., is Professor of Biology at Millersville University, Millersville, PA 17551.

Edwin L. Oxlade, Ph.D., lectures in the Science Department of Stranmillis College, Belfast, Northern Ireland, BT9 5DY.

Richard J. Rezba, Ph.D., is a professor in the Division of Teacher Education, Virginia Commonwealth University, Richmond, VA 23284.

A.J. Russo, Ph.D., is Associate Professor of Biology at Mt. St. Mary's College, Emmitsburg, MD 21727.

David J. Schimpf, Ph.D., is Associate Professor of Biology, University of Minnesota, Department of Biology, Duluth, MN 55812-2496.

William H. Sharp, Ph.D., is Assistant Professor of Biology, University of Lethbridge, Lethbridge, AB, Canada T1K 3M4.

Brian R. Shmaefsky, Ph.D., is Professor of Biotechnology, Kingwood College, 20000 Kingwood Drive, Kingwood, TX 77339.

Linda Sigismondi, Ph.D., is Assistant Professor of Biology at the University of Rio Grande, Rio Grande, OH 45674.

Steven L. Stephenson, Ph.D., is Professor of Biology at Fairmont State College, Fairmont, WV 26554.

Robert Tatina, Ph.D., is Professor of Biology, Dakota Wesleyan University, 1200 W. University Avenue, Mitchell, SD 57301.

Darrell S. Vodopich, Ph.D., is Associate Professor of Biology at Baylor University, Waco, TX 76798.

Robert P. Williams, Ph.D., taught and conducted research in microbiology for 37 years at Baylor College of Medicine, Texas Medical Center, Waco, TX. Dr. Williams passed away in 1993.

William J. Yurkiewicz, Ph.D., is Professor of Biology at Millersville University, Roddy Science Cntr., Millersville, PA 17551.

Index

Allelopathy	142
Allium cepa	19
Antidiuretic hormone (ADH)	125
Anthocyanin	19, 77
Aquatic ecosystems	154
Autotropism	92
Bacteria	25, 32
Bark	41
Beet (*Beta vulgaris*)	14
Beta vulgaris	14
Betacyanin	15
Betalains	78
CAM	81
Chlorophyll	98
Chromatography	98
Crassulacean Acid Metabolism (CAM)	81
Dandelion (*Taraxacum officinale*)	90
Daphnia	138
Data analysis	184
Dilutions	182
DNA	8, 25, 28
Duckweed (*Lemna*)	164
E. coli	8
Economics	136
Ecosystems	145
Enzymes	22
Fermentation	11
Food poisoning	128
Gibberellic acid (GA)	108
Gravitropism	90
Hardy-Weinberg principle	171
Helianthus annus	76
Hormone	107, 109, 125
IAA	109
Lemna	164
Lichens	160
Membranes	14
Microbes	69
Milk	32
Natural selection	150
Niche	136
Nitrogen fixation	47
Onion (*Allium cepa*)	19
Osmosis	19
pH	13, 22, 69, 77, 83
Photosynthesis	81
Pigments	15, 19, 77, 78, 98
Plasmolysis	20
Pneumococcus	8
Pollen	104
Pollution	160
Population genetics	171
Population growth	164
Pregnancy	126
Rhizobium	49
Saccharomyces cerevisiae	65
Sea urchin	118
Slime molds	41
Spectrophotometry	17
Stress	14
Sunflower (*Helianthus annuus*)	76
Taraxacum officinale	90
Tetrahymena	58
Transformation	8
Transpiration	76, 87
Tropisms	90, 92
Ultraviolet radiation	65
Urine	114, 125
Water flea	138
Water regulation	58
Yeast	23, 65